Puzzles, Problems and Enigmas

Philosophers have measur'd mountains,
Fathom'd depths of seas, of states, and kings,
Walk'd with a staffe to heav'n, and traced fountains:
But there are two vast, spacious things,
The which to measure it doth more behove:
Yet few there are that sound them; Sinne and Love.

George Herbert

JOHN ZIMAN, FRS

Puzzles, Problems and Enigmas

OCCASIONAL PIECES ON

THE HUMAN ASPECTS OF SCIENCE

CAMBRIDGE UNIVERSITY PRESS

Cambridge

London New York New Rochelle

Melbourne Sydney

CAMBRIDGE UNIVERSITY PRESS
Cambridge, New York, Melbourne, Madrid, Cape Town, Singapore,
São Paulo, Delhi, Dubai, Tokyo

Cambridge University Press
The Edinburgh Building, Cambridge CB2 8RU, UK

Published in the United States of America by Cambridge University Press, New York

www.cambridge.org
Information on this title: www.cambridge.org/9780521136341

First published 1981
This digitally printed version 2010

A catalogue record for this publication is available from the British Library

ISBN 978-0-521-23659-1 Hardback
ISBN 978-0-521-13634-1 Paperback

CONTENTS

APOLOGY

It would be idle to pretend that the book reviews, radio talks, public lectures and other pieces reprinted in this volume make a connected discourse. All that they have in common is the broad subject of modern science, seen as a multifarious human activity. This subject is now much discussed in many different circles, but I am not sure that all the notions introduced in these pieces have been totally discredited, or have yet become totally familiar. This is the only motive, beyond the normal vanity of an author, for having them republished.

In my first thoughts about this volume, I had it in mind to select various passages from the original articles, and to link them into an ostensibly integrated work. But when I sat down to the task, I realized that I would then be committing myself to a substantially new book; almost every sentence would have had to be rewritten to fit smoothly into the flow of the argument. Although such a work might be much more satisfactory than what is offered here, I could not feel that the labour would be well applied: better to start afresh upon a narrower and more modest theme than to attempt to fashion whole cloth out of such a patchwork.

The style of thought and exposition that is appropriate to each such 'occasion' is not suited to a larger, more deliberate theme. If these pieces are adequate as essentially journalistic sallies, this is because they were composed individually and read best as separate pieces. Indolence and literary vanity combined to instruct me to leave each piece in its original form, omitting only a few short passages dealing specifically with the books being reviewed. I have not attempted even to embed the various articles into a more comprehensive structure of editorial notes and explanations; they are all written plainly enough to be self-explanatory, and are sufficiently robust to stand up for themselves in open debate.

The advantage of the occasional piece, stimulated by some good or bad book or by the invitation to lecture to some special audience, is that it offers the opportunity to explore some particular aspect of the larger subject unhampered by an obligation to make one's findings seem to fit into a consistent overall scheme. It would be difficult, for example, to conceive a unified book that would include such interesting but disjoint topics as the philosophy of theoretical physics, technical problems of the dissemination of scientific information and what James Watson said about Sir Lawrence Bragg in *The Double Helix.* Yet each of these is a significant aspect of the scientific life, worth study in its own right. The best I could do was to sort the various pieces into rough categories, indicating their approximate relevance. This ordering, although not obviously unique, is at least more rational than a chronological sequence by date of original publication.

As might be expected, there is quite a lot of overlap between various pieces. In particular, I found myself advocating, again and again, the idea of science as a cooperative activity, first put forward in 1960 ('Science is Social', Chapter 6) and later set out at length in *Public Knowledge* (1968) and *Reliable Knowledge* (1978), both published by the Cambridge University Press. If the perceptive reader diagnoses mild monomania on this topic he is quite right; for me, at least, this idea is the key to many of the intellectual, psychological and social phenomena about which this book is mainly concerned, and I shall continue to press this point of view until it has been shown to be decisively wrong. Since academic history, philosophy and sociology have been moving in much this direction during the period since that article was written, I am not too fearful now of being put away for not being of sound mind.

Another major theme – the professionalization and 'industrialization' of science in the past half century – was also expressed quite forcibly in 1960 ('Scientists: Gentlemen or Players?', Chapter 15) and has dominated my more 'political' writing ever since. In these essays I have taken for granted the evidence for this historical transformation; a more direct account of the actual process will be found in *The Force of Knowledge* (Cambridge University Press, 1976). Here again, scholarly and lay opinion have flowed in much the same channels, so that what is reprinted here now seems quite tame and conventional by comparison with the views of some of our contemporary intellectual *avant garde.*

Rereading these pieces one by one, I did not find myself in total disagreement with my own opinions of some years before. This is disconcerting; in the intellectual sphere, transformation by reassessment is a higher virtue than consistency. What made me even more uneasy is the tendentiousness of much of the writing; such confident opinions expressed so positively, with so little hard evidence, on so many subtle questions. The collection, as a whole, is too journalistic, and lacks scholarly precision.

But this is not, I would submit, entirely my own fault. As suggested at several points in this book, the sociology of science is fragmentary and quite inadequate to its potential scope. On many of the topics here discussed, there is very little hard evidence available. In such circumstances, there is something to be said for raising problems, posing questions, and even positively asserting speculative theories, as a guide or a provocation to real scholarly efforts. My excuse for not having made such efforts myself is that almost all the material in this book was written in my 'spare' time, at home in the evenings or at weekends, on trains or in aircraft, away from the daily (if not specially onerous) duties of a full-time member of the academic staff of a distinguished British university.

I am grateful to various journals and other authorities for permission to republish these articles: all such sources are acknowledged as they occur in the text. Once again, it is a pleasure to thank my secretary, Mrs Lilian Murphy, for typing and retyping so many pages of wild and woolly manuscript, and for keeping tabs on the various articles as they were published. As for the conventional expression of appreciation of my wife and family: I know they really do understand and forgive that horrible itch for scribbling that has taken me down to my study, away from them all, for so many evenings over so many years.

Bristol, April 1980 J M Z

PART ONE

RESEARCH AS AN ART

1

Puzzles, Problems and Enigmas*

This is what research means for the individual scientist.

When dons discuss the research they are doing, they refer to it as 'their own work'. But the tone of voice they use is that of children talking about playtime.

Real scientific research is very like play. It is unguided, personal activity, perfectly serious for those taking part, drawing unsuspected imaginative forces from the inner being, and deeply satisfying. The sociological theory that it is performed to produce contributions to learned journals in exchange for professional recognition may not be quite nonsense – but it is totally irrelevant to the psychology of the researcher. The alternative theory that the scientist is merely the agent of economic forces and social classes in an inevitable historical process may also have some merit, taking a broad view, but again it does not explain the passion with which this strange activity is pursued. Science is not a job; it is an obsession. One knows of old scholars, full of fame and honours, who never can retire; their addiction is unconquerable, like that of an elderly duke who can never leave off fox-hunting no matter how dangerous the falls for his brittle bones.

What is so fascinating about it? To sit for hours over a microscope counting cells, or particle tracks, or spines on the eggs of flies! To worry oneself sleepless over an algebraic equation or the wiring of an electric circuit! To read and reread, to write, and rewrite, a half page of mysterious jargon that only 17 other people in the world will really understand! That is what scientists seem to do all day, and half the night; why don't they get themselves healthy, manly jobs, like piloting aeroplanes, or managing insurance companies, which would be more fun, more leisurely, and better paid?

Scientific research is solving puzzles. The pleasure to be got from it is

* BBC Radio broadcast, January 1972, as part of the science curriculum of the Open University.

the pleasure of the crossword or jig-saw addict. First the blank diagram, or the scatter of meaningless pieces; then an occasional tentative clue or the few pieces of the same colour that seem to fit together; next a period of frustration, going over and over the list of clues, or trying piece after piece in the most unlikely conjunctions; then – ah, the sweet joy of the word that completes a doubtful acrostic, or the section that springs to life as a tree, or a house or a pot of flowers; finally, the completion of the pattern, with clue after clue solved in rapid succession, or the last few pieces tumbling into place. By accepting the challenge, the tension, the concentration, the frustration, we heighten the pleasure of the moments of revelation. The more difficult the puzzle, the greater the tension – and so much greater the delights of solution.

Scientific puzzles are truly difficult. The solutions may take months, or years; sometimes we must leave them unsolved to tantalize our successors. Think of Fermat's last theorem, an apparently simple property of numbers which *he* said he could demonstrate. It was published exactly three centuries ago, and is still neither proven nor found to be false. Or recall the long mystery of the magnetism of the Earth, first studied by Gilbert who was court physician to Queen Elizabeth, and only recently resolved by the convective dynamo model. Such puzzles, indeed, are never quite solved; unlike a crossword or a jig-saw, they have no boundaries; chop off one head, and, hydra-like, a hundred new ones grin maliciously at us.

One of the charms of a scientific puzzle is that we are seldom trying to tackle it alone. All over the world we discover unknown colleagues, working away on the same part of the picture, themselves putting in a piece here and there, and showing their appreciation when we too are successful. Much of the pleasure to be got from science is this comradeship – the friendship of enthusiasts with the same peculiar hobby. But if we take our game too seriously, we can also be badly hurt if we are forestalled in a discovery or shown to be wrong. I must admit, as a jig-saw addict, that I can't bear another member of the family leaning over my shoulder trying to put in some pieces – especially if one or the other of us has fitted a section together incorrectly!

It is sometimes asserted that modern science is a rat-race – everyone trying to do the other fellow down. This does not seem to be an essential feature of research, except in very fashionable fields, where a competitive spirit is deliberately fostered, and prizes are awarded to those who seem to have won a race to complete a promising section of the puzzle. There are really quite adequate rewards for any competent research worker in the tussle with Nature itself. Intense scientific rivalry causes as much damage by emotional stress as it achieves a spur to effort.

The fact that science is a communal activity provides the only strict rule of the game: the answer that we offer to any scientific question must be good enough to convince our scientific colleagues. It is useless to pretend that we have a solution if we cannot demonstrate it clearly to other people and persuade them that it is true. In a very broad sense, this is the difference between a science and an art, such as painting or poetry, where our statements about the world may be as private and personal as we care to make them. The purpose of science is to create a body of *public* knowledge, not an impalpable, shifting scenario that could be a camel for one man and a weasel for another.

This rule imposes rigid constraints on the content and methodology of research; I would myself say that it defines the whole activity. But then a good tough rule like this, which prohibits the soft, easy, self-deceiving solutions, is just what we need to make a puzzle really interesting. At least it gives us a reasonable test, beyond mere fashionable acclaim, of whether we are anywhere near right in our answer.

This is one of the basic difficulties; the correct solution of today's scientific puzzle is not waiting for us on the back page of tomorrow's newspaper. We cannot look at the picture on the box, or turn to the list of answers at the back of the book. Genuine scientific puzzles are not at all like the conventional 'problems' and 'exercises' at the end of each chapter of a school textbook. By definition, no-one knows the answers in advance, and only time will tell whether an apparently convincing solution is the real truth. This is a feature of research that often distresses the student, who has become accustomed to the closed, convergent intellectual world of elementary basic science, where it seems inconceivable that every simple question should not have a simple definite answer.

Many grown-up research workers fail to escape from this psychological scaffolding. The puzzles that they attack are clearly defined, within the current scientific conventions of the day, and the methods for their solution are familiar and well-tried. Such puzzles are to be found repeatedly in the laborious and expensive process of technical development – the design and manufacture of machines for everyday use – but they are also a major part of academic science under normal circumstances. Not only do they provide pleasurable employment for a great many talented people, but their solution is necessary for scientific progress. It is a great error to suppose, for example, that classical physics is a closed system, already complete at the death of Lord Kelvin in 1907, and thereafter superseded by quantum theory. There are innumerable questions that can still be posed in the language of classical physics, such as the nature of the convective magnetohydrodynamic motion in the liquid core of the Earth, whose exact solution would be both difficult and immensely instructive. The fact that all the formal mathematical machinery for such a solution may be ready to hand does not tell us the answer by merely turning a handle. Imagination, intuition and intellectual invention are as valuable here as they would be in any attempt at a total theory of fundamental particles or of the history of the cosmos. The dry, formal language that is conventionally used to report scientific work of this kind is as inadequate as a marriage certificate is to describe a love affair. Nobody except a computer theorist pretends to solve interesting puzzles by a succession of purely mechanical deductions; the much more subtle problems of genuine science cannot conceivably yield to such procedures. Even within the most formal, academic and sterile branches of science, the logical apparatus of symbols, deductive geometry and digital computation are mere aids to imaginative thought of an intuitive kind.

The best evidence that research is very far from a mechanical process is the vast amount of error to be found in the archives of science. Another psychological shock for the well-brought-up, conformist science student is the discovery that three quarters of the scientific papers on any subject are wrong. The sources of error are manifold – experimental inaccuracy, mistakes of arithmetic, neglect of significant disturbing effects, theoretical misconceptions, etc. The work is done in full earnest – and yet it fails to be quite true or convincing. This does not mean that most scientists are fools; it means, quite simply, that research is always much more difficult than it seems, and that truth is a very stern master. The man who can always get things right is immensely respected – but we are forgiving of honest error, for we know the pitfalls on the way.

For this reason, puzzle-solving within the boundaries of an established scientific discipline is by no means uncreative. Although the main lines of general theory may be correct, the archives must be purged of innumerable minor errors and misunderstandings and augmented with a multitude of further observations. As with a great deal of practical, applied science and technology, immense satisfaction may be obtained from the mere exercise of specialized skills and experience – the delight of the master craftsman in his own handiwork. It is the wise scientist who knows himself well enough to be content with this level of achievement, rather than battering himself to pieces in the search for a 'breakthrough'. Indeed, this is the limit of intellectual independence permitted to most of those who take part in the vast experiments of Big Science, where a whole team must cooperate in the use of very expensive apparatus.

Yet science does not stand still. As puzzle after puzzle is solved, within some orthodox framework of theory and method, we begin to perceive new *problems* – questions that can be easily formulated but have no obvious answers. In my student days, a typical problem of physics might have been 'What is the nature of nuclear forces?' or 'Why are some metals superconducting?' A corresponding biological problem would have been 'What is the chemical mechanism of heredity?' These problems have now been solved – but they have been replaced by others such as 'What is the connexion between quantum theory and gravitation?' or 'What is the cause of biological aging?'

For most pure scientists, the attempt to solve problems is the real goal of research. It is the problems that make the subject exciting and challenging, carrying one away from the settled textbook regions to the unexplored frontiers of knowledge. Problem-solving is difficult indeed, for it demands great leaps of insight and imagination, but it also wins the big prizes and makes the big names. It is a mistake to suppose, however, that only speculative imagination contributes to the overall progress of science. Much academic research that seems, superficially, to be mere puzzle-solving is directed more subtly, towards some larger problem, perhaps by clearing away minor errors and uncertainties that seem to confuse the issue. The clever-clever guys, with their fertile inventive minds, must be balanced by the learned scholars, who use their intellectual powers to criticize and reject the wilder fantasies, and thus keep science from running away into nonsense.

The fascination of research for the mature scientist is this balance, this tension, between beautiful new conjectures and well-established fact – a tension in which one must keep one's mind open to every possibility, and yet never be blind to unanswerable objections. Medawar has called it 'The art of the soluble' – the skill to choose questions that are not trivial, and yet can be attacked and answered within a lifetime of effort. For each of us, this strategy must be a personal decision, based upon a shrewd estimate of our own capabilities, our own strength of will and of the current state of the art. The framework of choice is always changing, as new discoveries are made and new problems come to the surface. It is a sobering exercise to look back, perhaps no more than five or ten years, and to see how the whole climate of opinion has altered, so that old mountains of difficulty have now shrunk into molehills, and old fogs of mystery have evaporated into clear air.

A certain spirit of conservatism is inherent in all branches of scholarship, for the hard-won truths of the past must at all costs be preserved. The social system of the scientific community, by giving great weight to the authority of age and experience, often puts obstacles in the way of the revolutionary thinker – as, for example, in the case of the theory of continental drift, which was rejected for half a century by 'official' geology. It is easy enough, with the wisdom of hindsight, to see the follies of our ancestors. We know very well that the active scientist must be an anarchist at heart, ready to overthrow the old gang as soon as he doubts their credentials. But this is a counsel of perfection; the habits of thought, the mature experience, of a lifetime are not thus recklessly to be discarded.

For this reason, the highest honours of science are accorded to the revolutionary geniuses who do more than solve problems – they perceive that problems exist. The contribution of a Galileo, a Darwin or a Wegener may be described as the formulation and resolution of an *enigma* – a question concerning a deep and mysterious feature of existence, such as physical motion, biological speciation or the shape of the continents, that had scarcely been asked before. To dare to propose such problems for scientific investigation, they had to step out of their own mental skins; they had to look again through the eyes of childhood; they had to exercise uninhibited curiosity.

It is often said that real scientists are full of natural curiosity, but in our own highly organized scientific profession this talent is only cultivated in an institutional, collective manner. Without individual curiosity, there could have been no notion of an enigma; without an unhealthy interest in the enigmatic, there would be no scientific problems; only by the solution of problems have we acquired the technical power to solve the puzzles of our complex civilization.

2

Einstein*

He was the greatest; what does he mean to us now?

The traveller of independent mind eventually discovers, alas, that the popular sights – the Grand Canyon, the Sistine Chapel – deserve their reputation. It is well to be reminded that the most famous scientist of our times merited equally unstinted admiration. It was a pleasure and an inspiration to revisit this great intellectual monument with such civilized guides, to catch glimpses of this gentle genius at his creative task.

What does Einstein mean to us now? To the general public, he is a magician who uttered the formula by which matter could be turned into energy and who conjured away the reality of space and time. To the physics graduate, he is the most brilliant of all the professional star performers, contributing to every branch of modern physics – mechanics, electromagnetism, quantum theory, cosmology, statistical mechanics and the theory of solids – brief, lucid, modestly phrased papers that pierced to the heart of the mystery and made all things clear.

Only a few thousand experts are familiar with his greatest work, the general theory of relativity, which showed that gravitation could be represented as space–time geometry. It is a work of such physical, philosophical and mathematical originality and depth that it is doubtful whether anything like it could ever have been produced by any other man who has ever lived. Let us admit it: he *was* the greatest.

We admire also the lifestyle, so simple and modest, so kind and friendly, so courteous and good humoured, so liberal, wise and humane. Newton, as we are now informed, was arrogant, vain, suspicious and quarrelsome. It is difficult to imagine the character defects that will be discerned in our Einstein, 300 years from now. The failure of his first marriage seems no more than a mild accident of fortune. It was nothing to be ashamed of that he could not accept the probabilistic interpretation

* From a review of *Einstein* by Jeremy Bernstein (Fontana, 1973), and *Einstein: The Man and his Achievement,* edited by G. J. Whitrow (Dover, 1973), in the *New York Times Review,* September 23, 1973.

of quantum mechanics. Nor that he lavished his efforts for the last 30 years of life, without obvious success, on attempts to unify electromagnetism with gravitation. Like many other great artists, he had moved far beyond the tastes and style he had created for his contemporaries, and worked to satisfy only his own standards.

It is sad, nevertheless, that the 'age of Einstein' ended long before his death in 1955. Nazi persecution and exile in America broke into his life when he was 50, but were not so personally tragic for him as for several million other European Jews. He continued to the end as he had begun, dedicated to research, clear and concentrated in mind, undistracted by the world. But science itself moved away from Einstein.

Is he now, in Herbert Dingle's words, 'the popular idea of the typical scientist'? Can one honestly advise anyone to follow his example? How foolish not to take school seriously, not to strive to get a good degree! What scientific job will there be for the man who despises the lecture programmes and thesis advisers of graduate school, who fails to get a Ph.D.? How will he obtain access to the reprint exchanges, computer programs and scientific conferences as an assistant in a patent office? Will his papers be acceptable – without citations to all the current literature and unaccompanied by a bank draft to cover the page charges? How long will his promotion be deferred, as he fails for ten years to solve an apparently crazy problem? Where will he get the research grants to maintain his scientific reputation – if he cannot tell the research council what he will discover? What will they think of him in the nation's capital if he refuses, as a pacifist, to advise the Pentagon secretly on new weapon systems? Or, if he happens to be a Russian, will he be permitted to go to Israel even if he has done no secret work? One can even say that physics no longer has need of such intellects, that all the big problems have been solved, that all we are doing now is 'normal science,' filling in the details, exercising our minds with little puzzles, and devising more and more frivolous or destructive gadgets.

And yet, one day, who knows, there may arrive on my desk a scruffy handwritten manuscript, from a forgotten student, who got a Third and went off to become a lighthouse keeper (Einstein's ideal occupation!), explaining the mass spectrum of the elementary particles, or the nature of quasars, or the connexion between electromagnetism and gravitation. I shall be wise (and the world will be lucky) if I take a second look at it, and consider whether, beneath the fantastic notions, there may not lie a universe of truth.

3

Themata*

What were Einstein's ideas, and where did they come from?

Professor Holton illustrates one of his essays with a *colère,* by the French artist Arman, consisting of the shattered instruments of a string quartet; to symbolize the whole collection of articles reprinted as *Thematic Origins of Scientific Thought* he might have used one of Gian Piero Pirovano's imaginative representations of the chaos of things *before* they are made. The 'science' of the published paper, which the philosophers so earnestly analyse, is S_2; the chrysalis of an idea in the mind of the scholar is S_1, essentially private and yet the source and origin of public knowledge. What goes into the creation of a scientific theory? Can we make a plausible reconstruction of the process of intellectual discovery?

The task is more difficult than it looks. The public statement of a new theory is not intended as a confession or an autobiographical playback of the creative act. As Holton puts it, S_2 is projected onto the 'x–y plane' – the dimensions of empirical fact and logical theory. But S_1 extends deep into the 'z' dimension, where tacit principles and intuitive comprehension are equally valid. His purpose, expressed in characteristically lucid, refined and scholarly prose, is to discover the *themata* that have been called into action in that private world. In a series of detailed studies, mainly concentrated on the 'nascent moment' of 1905 when Albert Einstein conceived the theory of relativity, he shows how much can be achieved by very careful attention to biographical details, autobiographical comments and the substance of the published work itself. It is a delicate intellectual exercise, of particular fascination for anyone who is familiar with modern physics, illuminating the inner meaning of scientific knowledge.

Every critic knows that the reconstruction of the creative act of any great artist can never be better than provocative speculation. But theoretical physics is an art whose elements are not beyond reckoning. For all his

* From a review of *Thematic Origins of Scientific Thought: Kepler to Einstein* by Gerald Holton (Harvard University Press, 1973); paid for, but apparently not published, by the *New York Times Book Review*.

magnificent genius, Einstein produced what any first-rate theoretical physicist could conceive of having himself produced in the circumstances. The same can scarcely be said of, say, a Joyce, a Stravinsky, or a Picasso: *his* work of genius springs so completely from his own life history and particular circumstances that it could never have been produced except by just that person at just that moment. The fascination of Einstein's paper is the interaction between the objective situation – the physics of the day, the facts and theories to be explained or reconciled – and the subjective force of this young man of 26, his mind, his personality, his interests and his abilities. Even if the knot cannot be untied completely, we are not bound to slash wildly at it in desperation.

As usual, the conventional historical account of the physics textbooks is all wrong. If Einstein had been through a proper graduate school, he could have been made to study carefully all the papers of Poincaré and Lorentz – two absolutely first-class theoretical physicists who came very close to the relativistic equations but who failed to see the central issue. He would have known all about the Michelson–Morley experiment, which failed to show up the motion of the Earth relative to the ether, and would perhaps have conjectured that the velocity of light must be independent of the motion of the observer. His research supervisor would then have insisted that his thesis should contain a careful analysis of this and other observations, showing in detail how all the anomalies could thus be explained.

In fact, as a schoolboy, ten years before, Einstein had come intuitively to the conclusion that the idea of 'catching up with light' was physically absurd. The textbook from which he later learnt electromagnetic theory happened to emphasize a peculiar feature of the calculations of the current induced in an electrical conductor moving in a magnetic field. He was also strongly influenced by the philosophical principles (which he later turned against) of Ernst Mach, which argued against the introduction of unobservable entities such as absolute space. The sources of his theory were not, then, the 'paradigms' of classical physics as taught by the professors of his day, but the deeper 'themata' of intellectual coherence, the uniformity of nature, mathematical invariance, simplicity and symmetry. Once he had projected these 'z-dimensional' elements into a coherent set of principles and equations in the 'x–y plane' there was no more to say; Michelson–Morley and all the rest follow automatically, without special analysis, just as the theorem of Pythagoras follows from the axioms of Euclid: QED.

For the general reader, the most interesting essay must be Holton's final attempt to understand this particular type of genius. He draws attention to Einstein's powerful visual imagination – stimulated and encouraged, perhaps, by a short period, when he was about 16, at the Aarau school, which was run on 'Pestalozzi' principles. There is no doubt, in my mind, that theoretical physics demands this type of imagery in the highest order; the manipulation of symbols is sterile without an intuitive grasp, beyond verbal or algebraic representation, of what must really be going on. But the essential talent is the ability *not* to be constrained within any particular method or thema. Every scientific problem, every challenge to the intelligence, calls for its own technique of solution. For every thema there is an equally powerful 'anti-thema': continuity faces discreteness; permanence faces change; order opposes chaos; phenomena challenge rational ideals; the romantic plunge into the unknown is no more valid than the return to classical purity. In a great mind, thema and anti-thema are both present, and both available for duty; neither is permanently subordinated to the other by some petty principle of non-contradiction.

Could there be some value in cataloguing and commenting on the sublime themata of science and other scholarly pursuits? Holton emphasizes their small number, and ever-living presence in philosophical thought, from the Greeks onward; like Indian gods, they sustain the world by the continued tension of their opposition. Are they the archetypes of Jung; are they built inextricably into our schemes of thought by the language that we use; or are they imprinted upon us, by the world of reality, in childhood experience? They cannot belong entirely to the private realm of S_1, for the whole study would be meaningless if we could not partially share them with Einstein himself. Indeed, the suggestion that S_2 science lies entirely in the x–y plane is too strong. In our own response to Einstein's paper on relativity, we are not completely persuaded by the purely logical or empirical arguments. In so far as we have been trained to 'think physically', we recognize and resonate to just those thematic elements that Einstein sang so sweetly for us.

One of the tasks of public science, over the centuries, is to refine and reformulate the old themata, giving them new meanings and new power. The permanence of Parmenides reaches ultimate perfection in the space–time continuum of general relativity; the continual flux of Heracleitos is purified into the zero-point motion and virtual transitions of quantum mechanics. In his essay on 'The Roots of Complementarity' Holton even suggests that Niels Bohr had created an entirely new thema. The technical reconciliation of the apparently contradictory concepts of wave and particle in the interpretation of quantum phenomena is of extraordinary philosophical depth and generality. I would go even further, and elevate it into a 'meta-thema' – a metaphor for the acceptance (backs to the wall, almost) of the essential contradictoriness and imperfection of all our attempted insights into the mysterious world we inhabit. Out of materialistic science comes this spiritual solace, closing the circle from rationalism to Zen.

4

The Quest of the Golden Helix*

The scientific life, in reality and as myth.

As Sydney Smith once remarked – only half in jest – 'It's a great mistake to read a book before reviewing it. It prejudices one so!'

When I heard that James Watson had written a book about the discovery of the structure of DNA, for which, along with Francis Crick and Maurice Wilkins, he had won the Nobel Prize, I soon knew what to think. It had been refused publication by the Harvard University Press; it had been serialized by the *Atlantic Monthly;* and many of the persons mentioned in it had objected to its appearance. Public gossip, in the Hollywood manner, had come to the scientific community.

It was only to be expected. Science becomes bigger and more important. Leading Scientists become even more Top People. The lime-light of Mass Culture is projected onto them by Press and television. A star system grows up. Learned professors of astrobotany and molecular chromatonomy expect notes to be taken of their breakfast-table platitudes – as well as salaries of $100 000 a year. Vanity feeds on publicity and flattery, but is not sated. The mass never tires of news about the few names that it knows – but news must be 'interesting', therefore scandalous. The customary reticence and privacy of the scholarly life is violated. The vain, the exhibitionist, the unscrupulously ambitious, profit at the expense of modest conscientious citizens. The Republic of Learning is corrupted at its very heart.

I need not say how disastrous this would be. A top scientist is *not* a paid entertainer, whose audience is enlarged by any publicity, however discreditable. The idealized conventions of scientific journals make the scientist appear impersonal, almost anonymous, so that his experiments and theories may speak out for themselves. In practice, other scientists praise and honour him for his achievements, and his name at the head of

* BBC Radio broadcast, 20 May, 1968, reviewing *The Double Helix* by James Watson (London: Weidenfeld & Nicolson, 1968).

an article becomes some guarantee of its validity. The worship of star performers is therefore very dangerous, for it makes too much of the words of those who are already successful, and encourages the desire for personal acclaim rather than for the unveiling of the truth. Science is utterly dependent on the professional honesty and integrity of every active research worker. Public gossip – necessarily scandalous, false, malicious, wounding – would inhibit genuine argument, terrify the timid into orthodoxy, and destroy the web of mutual confidence supporting the whole system.

It was easy, therefore, to know what to say about *The Double Helix.* However interesting such a memoir might be for the historian of science, there could be no justification for a vulgar exposé of the characters and activities of living individuals against their wishes. I do not think I am just being stuffy. The laws of slander and libel have not been repealed. Gossip-mongering about the private life of a scholar could be as destructive of his self-esteem and professional standing as it would be for a doctor, lawyer or company director. There can be no claim here for privilege in the public interest; on the contrary, science is a profession where 'the cult of personality' is entirely antisocial.

Alas for my splendid sermon; the book itself belies its own reputation. It is compelling reading – not for scandalous revelations but as a human story.

In fact it is an accomplished work of art, with the simple charm of an old-fashioned fairy tale. 'Honest Jim', a poor, clumsy, ignorant peasant lad sets out to make his fortune. He learns of a magic treasure – the structure of DNA – and resolves to find it. After various adventures he arrives in Cambridge, where the Cavendish castle is governed by old Sir Lawrence. In the kitchen he befriends Francis, a scullion, who also dreams of knightly deeds, and practises at jousting with a broom handle, until sent back to scouring pots by grumpy old Sir Lawrence. But secretly Jim and Francis continue their search, with the aid of the sad squire Maurice and his encaged dragon Rosalind. In the end, they defeat the wily old Sir Linus, and discover the golden helix. Sir Lawrence is delighted, the Baron of Todd knights Jim and Francis on the spot, and even the dragon Rosalind turns out to have been a fairy godmother under a spell. Who are all the beautiful foreign girls that Jim keeps meeting? Since they are quite clearly not damsels in distress, I'm inclined to put them down as temptresses, sent by Sir Linus himself to distract our hero from his quest!

Innocence, clownishness, grumpiness, dark plots, persistence, cunning, glorious success – these are the elements of the story. But although it has the emotional pattern of a fairy tale, it is not fictitious. The episodes described with such verisimilitude occurred some 15 years ago, but Dr Watson has drawn heavily on his weekly letters to his parents in America to reconstruct the course of events.

I was not myself at the Cavendish at the time, but many of the persons in the tale are known to me, and I do not find their reported actions quite out of character. Sometimes innuendo verges on injustice – but perhaps that is how these strangely cold British scientists did seem to the brash, rather too pushing, young man that the author makes of himself. 'Those who figure in the book must read it in a very forgiving spirit' says Sir Lawrence Bragg in a preface – thereby, in one sentence, turning upside down the picture that Dr Watson had drawn of him. No, this is not a scandalous work, for it lacks malice. To tell the truth, I don't think the author knows or cares enough about other people's feelings to want to wound them. His sublime egotism is ungentlemanly, but innocent.

He says he did not set out to write a definitive history. Some serious-minded student who took no part in the events themselves will have to poke around in old files, check dates, try to read laboratory notebooks, weigh one bit of evidence against another and finally try to tell us what actually happened. *The Double Helix* reads very plausibly, but it could only be one version of a very complicated chain of events.

It is plausible, because it has an authenticity that is quite rare in such memoirs. It is no good going to the scientific journals for the truth about how a discovery was made; that would be as simple-minded as expecting to find out how a battle was won by reading the official despatches. Scientific papers are deliberately contrived to give a necessary and logical appearance to what actually began in darkness and was fought largely by accident to an unsuspected conclusion. Even in conversation, letters and informal seminars we seldom lay bare all our thoughts.

And as we dream up yet another interpretation of the phenomena, it occupies the same region of the mind as its predecessor, which is driven into oblivion. I can't recall exactly what I thought ten years ago about various scientific problems that have since been solved, any more than I can recall the appearance of my children as babies. I therefore admire the skill with which Dr Watson has consulted his letters and reconstructed the errors, misunderstandings, fallacies and reticences of successive phases of his research.

Above all, he tells us quite honestly how he felt on each significant occasion. Good scientists are often rather reserved, inhibited fellows, with icy layers about their inner thoughts. Their training in impersonality and abstraction makes it difficult for them to confess, in the spirit of Boswell or Rousseau, either shame or pride – why, indeed, should they? But they do have feelings, none the less. Let me quote the episode of Chapter 13. Crick and Watson have come to the conclusion that DNA is just a *triple* helix. Agog with excitement, they invite Maurice Wilkins and Rosalind Franklin down from London, to hear about the brilliant new idea. The X-ray experts come – and very quickly it becomes clear that the theoretical model will not fit the observations. 'After lunch,' Watson tells us, 'the situation did not improve when we got back to the lab. Francis did not want to surrender immediately, so he went through some of the actual details of how we went about the model-building. Nevertheless, he quickly lost heart when it became apparent that I was the only one joining the conversation. All its glamour had vanished, and the crudely improvised phosphorus atoms gave no hint that they would ever fit into something of value. Then when Maurice mentioned that, if they moved with haste, the bus might enable them to get the 3.40 train to Liverpool Street Station, we quickly said goodbye.'

Yes, the feeling is all too familiar. 'Ah well – back to the drawing board!'

Again, conventional courtesy does not prevent him from describing the tensions that arise between many of the characters in his drama. The relationship between Bragg and Crick exemplifies the problem facing the experienced professor with a brilliant but erratic research student. The 'verdict', that Crick should drop his attempts to unravel DNA and get back to his thesis task of investigating the ways that hemoglobin crystals shrink when they are placed in salt solutions of different density, was not as stupid and unjust as it may have seemed to our young heroes. To establish himself as a reputable scientist, the student must get his Ph.D. and publish a few sound papers. This is not because a Ph.D. is an Open Sesame magic formula, or even a certificate of professional competence. It is because science advances cautiously along a wire stretched between originality and criticism. The would-be scholar must feel this tension for himself, by carrying out a very careful piece of research in which all possible objections have been met and overcome. The more brilliant and original his theories, the more he needs critical self-discipline. Many a bright lad has been ruined by the indulgence of his research supervisor, and never learned to correct his own errors before publishing them. I guess that Bragg's decision was thoroughly justified, even though it did not endear him to this loud-voiced, bouncing, almost middle-aged student who *would* insist that he knew the answers to everybody's problems but his own. Poor Francis! Poor Sir Lawrence!

As it turned out, Bragg did not seem to have much power to enforce his decision, for Crick continued research on DNA none the less. This is, of course, as it should be. A large laboratory is quite a complicated organization, which has to be run in an orderly manner – but a research student must never be treated as a slave. His apprenticeship is to the art of becoming a 'self-winding' research worker, where he must take all the decisions for himself anyway. Watson, as a post-doctoral fellow, or research assistant, or whatever his status, was even more free to come and go at all hours, work on any topic he wished, and generally behave like an aristocrat or a tramp. In fact, as the book makes quite clear, his own zeal drove him far harder than any whips or scorpions.

But it isn't easy to preserve this atmosphere of freedom for young scientists, as laboratory equipment becomes more complex and as more and more money is needed for men and machinery. The story of Watson's fellowship is instructive. Without authority he moved from Copenhagen to Cambridge. I quote: 'Only in late January did my suspense end, with the arrival of a letter from Washington: I was sacked. The letter quoted the section of the fellowship award stating that the fellowship was valid only for work in the designated institution. My violation of this provision gave them no choice but to revoke the award.'

'The second paragraph gave the news that I had been awarded a completely new fellowship. I was not, however, to be let off merely with the long period of uncertainty. The second fellowship was not for the customary twelve-month period but explicitly terminated after eight months, in the middle of May. My real punishment in not following the Board's advice and going to Stockholm was a thousand dollars!'

It is hard work, indeed, for bureaucrats and cost accountants, having to deal with raw genius; they never seem able to bend the rules far enough to seem generous!

Yes, even fairy tales tell us truth about life – about poverty and riches, good and bad fortune, courage, loyalty and treason, love and death. Against the Cambridge backdrop of shabby lodgings and beautiful gardens, boring dons and clever students, dreadful meals and wild parties, *The Double Helix* tells us much of the truth about high pure science. The racy style, the puzzle itself, the psychological conflicts, the mysteriously irrelevant details about food and girls, compel our attention. It is the ideal school prize for every aspiring Nobel Laureate. It will make its own myth.

And that is more worrying than the gossip; for this book depicts science as a rat-race.

I admire the nonchalant self-confidence with which Dr Watson decided that he would get more from Cambridge than Copenhagen; not many British scientists, nursed through a succession of scholarships, studentships and fellowships, would have taken this risk of being cut off without a penny. I admire the single-mindedness of his attack on his problem, his intellectual vitality, and his youthful inability to admit defeat. These are the qualities for success, in science as in any other trade.

But there is a touch of ruthlessness and egotism about him that is less admirable. The 'romance of discovery' has degenerated into something closer to gang warfare. In the end, the search for an explanation of one of the most important phenomena in biology became merely a competition to be first with the correct solution of a difficult puzzle. Other scientists appear as either reluctant collaborators or potential rivals in the race; for Dr Watson they seem only to exist in relation to his own personal bid for glory.

As this book makes quite evident, the full story of the elucidation of the biochemical mechanism of heredity involved a whole host of other people. An idea was picked up here, a fact from there. Pauling's alpha helix, Chargaff's rules about base pairings – above all, the fantastic expertise of Bragg's school of biochemical crystallography – these were all essential to the final success. This was no case of Jack the Giant Killer, doing great deeds all on his lonesome. Crick and Watson were just the brave young sergeants who, after a mighty assault, finally planted the flag upon the summit of the citadel. By the time that they ventured into battle, victory was certain; it was largely chance that put the symbol of it into their particular hands.

The Nobel Prize, by honouring deeds of the mind and spirit, is a splendid institution. But as the sign of absolute excellence in science, and as a focus for personal ambition, it is a false god. The experts themselves will tell you how many superb scientists have been passed over because their discoveries were not 'important' to the view of the day, or because they were only stepping stones in a very long investigation by many different research workers. Those whom we should most honour are the pioneers, who glimpse the potentialities of an unexplored field, or who have the courage to look upon old problems in an entirely new light.

An early craving for personal fame can carry one into the heart of an existing, highly developed, fashionable discipline, to contend ferociously for the answer to the 'exciting' question that everybody is already asking. The supreme achievement is to ask the question in the first place.

Dr Watson's view of science is altogether too competitive. He takes for granted an atmosphere of jealousy and suspicion, which springs as much from his own character as from the realities of the scientific life. He does not see that it is a *cooperative* enterprise, in which the enemy is ignorance, not the chap in the other laboratory.

The fairy-tale morality may be good enough as a guide to action at the tender age of 23, but it is the wrong mythology for a long career. I believe that the proper ideal for the scientist is not the hero – the knight at arms – but the craftsman, the artist, the poet. The much-abused adjective 'creative' is correct in spirit. We do better to strive, not for 'success', but for perfection. I do not deny the tensions and conflicts that must arise because scientific excellence can only be judged by other scientists, whose duty it is to criticize and question all our efforts; the purpose of the ancient courtesies and reticences is to keep these conflicts under control. The hardest lesson for the young scientist is to bear his envy in silence, and to appear to rejoice in the achievements of his rivals. In opposition to the tone of this whole book, I would quote a remark of Marcel Proust – '. . . it is not the desire to become famous, but the habit of being laborious, that enables us to produce a finished work' – or Benjamin Jowett's humble advice – 'The way to get things done is not to mind who gets the credit for doing them!'

5

There were Giants in those Days*

A romantic view of science is not yet out of date.

Microbe Hunters by Paul de Kruif was first published in 1927; a reprint (July 1933) in Jonathan Cape's *Life and Letters Series* must have come to me when I was about ten. The binding is tattered from innumerable rereadings during the next few years of my life. As I read it once more, the other day, it spoke with the happily remembered voice of an old, old friend, uncrushed by time and change.

There was Leeuwenhoek, squinting at the animalcules that he had discovered through his tiny droplets of glass; there was Spallanzani, despatching his opponents with rapier-sharp experiments disproving spontaneous generation; Pasteur, with transcendental genius conquering the maladies of silkworms, beer – and rabies; Koch, all German doggedness, putting anthrax into its place; Metchnikoff, the mad Russian, discovering leucocytes and teaching salvation by yogurt the way Linus Pauling preaches now for vitamin C; and all the rest of them. The characters of the drama are as alive in my mind and heart as when I first met them, 40 years ago.

It is strange, I suppose, that a putative mathematician or physicist should have been so moved by, and still feel so obliged to, a book about bacteriology. But that sort of science was easier to grasp and to appreciate, as a child, than the more sophisticated triumphs of Faraday and Maxwell. It was the image of the scientific life that counted, not the particular subject. Perhaps it helped that this was pre-eminently 'good' science, devoted to the conquest of disease.

And it is, after all, a splendid book, simple, sure-footed and vivid in style and conception. Romantic, no doubt – but that is the proper tone in which to delight and enlighten a child. How can we face the contradictions and disappointments of life without a foundation of clear morality and remembered allegories of virtue? In fact, *Microbe Hunters* with-

* Article in *New Scientist*, 76, 795 (1977).

stands inspection by a more weary cynical eye; what it told me about the scientific life has not been falsified by my own experience, whether sweeter or more bitter.

First of all, it said, science is intensely exciting. There is no life like it for those who enjoy that sort of thing. The striving to understand, the passion to prove, the glory of discovery are the equal of all the strivings, the passions, the glories of knightly combat or other heroic deeds. The rewards were both private – the personal satisfaction of an enigma resolved – and public – acclaim for a benefit to mankind. The metaphorical joys of 'fighting the good fight' come naturally to any small boy; indeed, they were by no means quenched in the scientific heroes themselves, whose pugnacity to one another must have reminded me of those Norse Gods for whom I had also fallen in more childish reading. Who is to doubt that research demands the dedication and passion that can only come from sheer delight in intellectual feats of arms, with a distant vision of a Holy Grail of Truth to justify such a strenuous vocation. Mr de Kruif's heroes were just such champions to exemplify and arouse these emotions.

On the other hand, they were also very much human beings. There was no hiding the chicanery, the vanity, the paltry patriotism, the rivalry, the carelessness and many other defects of character that those nineteenth-century microbiologists could show in their lives and works. Of course, it is easy, in writing the history of science, to make the opponents of new discoveries seem pompous reactionaries; but the revolutionaries themselves were often quite impossible people whose only saving grace was that they happened to be right. Neither the revelations of *The Double Helix,* nor any personal experiences of academic folly or malice, could come as a complete surprise when one recalled how Pasteur and Koch had behaved to one another in those blissful days when even the physicists were not supposed to have 'known sin'.

How true was it as history? I have read very little serious scholarly work on the 'Pasteur era', so I have no idea – and would hate to be disillusioned on the subject. But *Microbe Hunters* was, in fact, the inspiration of an excellent BBC TV series in which the major episodes were re-enacted, and in the accompanying book (*Microbes and Men,* 1974) Robert Reid tells the same story as if that was how it really happened. Nowadays, with our gnawing doubts about the overwhelming beneficence of science, we might not go into quite such raptures as Paul de Kruif allowed himself from time to time – although I noted that he was as sceptical about the value of diphtheria antitoxin as we might be now about, say, chemotherapy of schizophrenia.

If, however, I feel a little nostalgia for a youthful time when that book could be reread half a dozen times in innocent anticipation of the real life to come, it is only that science itself has changed. The era of single combat, with sword and lance, microscope and scalpel, has passed, and we now fight for truth and health with Napoleonic big battalions of research teams, and the extravagant weapons systems of Big Science. Nevertheless, in many respects, private and public, it retains the excitement, the challenges, the human drama of a heroic age that deserves thus to be recalled.

PART TWO

INTELLECTUAL AND PHILOSOPHICAL ISSUES

6

Science is Social*

Science is best characterized as a cooperative activity, aimed at establishing the widest possible consensus. This concept underlies many of the other papers in this collection, and was later enlarged into a book, Public Knowledge, *published in 1968* (Cambridge University Press).

I am going to try to justify the following proposition: 'A scientific truth is a statement that has been publicly accepted by the experts.' This proposition may seem contrary to the spirit of free inquiry but I am not trying to accuse science of orthodoxy or authoritarianism; nor do I pretend that it is a complete self-contained definition. But I hope to show that, if it is suitably interpreted, this paradoxical proposition can be defended.

Let me start with an entirely practical question: how do I actually set about finding out what we 'know' scientifically about a particular matter? For example, I want to know how much light is reflected from a highly polished metal surface. This is a well-defined scientific problem and there must exist much scientific knowledge about it. So I do not spend five years and £10 000 setting up apparatus to measure it; I go into a library and browse. First I consult standard textbooks on metals, where I find certain general statements – perhaps an account of the theories and a table of experimental data. These are not sufficient, but the text is liberally sprinkled with footnotes and references. I pull down various volumes of the important scientific journals, and seek out the original papers in which the measurements were first recorded and the theories first formulated. In other special journals there will be abstracts and indexes which will help me to make a complete survey and to read every scientific paper that has ever been published on the subject. Short of writing to the authors for further private information about their experiments, there is nothing I can do but accept these published papers as the sum of our scientific knowledge on this topic.

The trouble is that these papers will not be consistent with one another. In one I shall read that the absorptivity of copper at a wavelength of ten microns is 0.6 per cent; another author tells me that it is only 0.4 per

* BBC Radio broadcast, reprinted in *The Listener* 64, 251–2 (1960).

cent. Smith says that this is due to a surface contribution from the anomalous skin effect; Brown claims that it is a volume effect arising from the anisotropy of the relaxation time of the conduction electrons. My professional training has taught me the meaning of these conflicting statements but it cannot tell me which (on the face of it) is true. Without a detailed experimental and theoretical investigation myself, I do not know whom to believe.

The point I am making is that scientific knowledge at this stage is usually chaotic, uncertain and inconsistent. Only afterwards, when we look back a few years, do we see clearly which experiments were sound and which contained errors; which theories were true and which were false. Yet when each paper was published all the traditional prescriptions of the 'scientific method' had been obeyed; the experiments had been done, the measurements checked, the theory formulated, the prediction confirmed. What more need have been done to arrive at the scientific truth?

What is missing from the standard descriptions of science is an account of the processes that follow publication of original work. Let me go back to the textbook I first consulted. If it is a good one, the author will have attempted to reconcile the discordant results. There will be a suggestion that one of the measurements was inaccurate because the surface was not properly polished. Or it will be pointed out that Smith's theory ignores an important factor. There will be references to other papers that offer decisions between alternatives, and if the point is not settled the author will say, honestly, that there is a 'discrepancy' which has not yet been explained.

In other words, there is, in science, a continuous activity of criticism, reassessment, and re-evaluation. A scientific paper is seldom the report of an isolated inquiry; it is deeply embedded among all the other papers on the subject. Its content does not become a scientific truth until it has passed through the furnace of critical appraisal and has been accepted by all (or nearly all) the other workers in the field. It is then incorporated in the 'canon' of the science, and will appear as a 'fact', without special comment, in all the books. This is what my proposition means: 'A scientific truth is a statement that has been publicly accepted by the experts.'

This is a loose and incomplete definition; we must decide, for example, what we mean by 'the experts'. I suppose that an expert, in science, is somebody who can make a scientific statement which is intelligible to other scientists, just as a man may be said to be expert at speaking French if he can make noises that are intelligible to a well-educated Frenchman.

This sounds like a circular definition of a closed mutual-admiration society, but we know, in fact, that this group of experts is always open to those who have the will to learn about the science and the wit to make sense of it. Moreover, we create social institutions such as the Royal Society, whose role is to recognize the most expert scientists and give their opinions special weight in the assessment of new scientific ideas. But I would agree that one of the difficulties in the establishment of an entirely new science – sociology, for example – is the creation of a group of reliable experts who can be trusted to give fair consideration to all new work.

Then the phrase 'publicly accepted' needs amplification. I do not mean that there must have been, at any moment, a formal decree of orthodoxy. But the very fact of publication in a reputable scientific journal represents qualified approval. Every paper is 'refereed' by experts, and may be rejected if there is some gross error or omission. The next stage may be comment in a 'review article' or book, where all the recent papers in a particular field are discussed and criticized. Even then, it may be several more years before it becomes apparent that all criticism has been stilled and the new fact or theory has become part of the standard mental equipment of every expert in the subject. Sometimes there has been a gap of many decades between the first formulation of an idea and its eventual complete justification.

There is a practical problem of making science work as efficiently as possible. I believe that in our training of scientists we put far too little emphasis on the role of communication. It is not enough to make a measurement or a calculation; one must also express the results lucidly, logically, and accurately, making it perfectly clear to the reader what it is one has done. How can the other experts be convinced if our statement is equivocal or obscure? As in the law courts, an ambiguous statement is 'void for uncertainty'. The special vocabulary and complex phraseology of scientific papers was not invented just to make science into an esoteric cult; it has the deliberate function of ensuring precision. But it is only too easy to string together jargon phrases and hope that some sort of meaning will emerge. The act of 'writing up one's results' is not a tiresone chore and a distraction from 'real' research; it is a highly creative activity, for we do not always know exactly what we have discovered until we set it down on paper for others to read.

Nor do we emphasize sufficiently the role of review articles and books. Everyone knows the influence of a standard treatise – it will set the tone and establish a point of view for a generation of students and scholars. Yet the writing of books is not very highly esteemed by comparison with 'original' research, and tends to be neglected. Too often nowadays a thick volume with an important title turns out to be the report of a conference or symposium – a mere collection of separate papers by a number of different authors all jumping on their hobby horses and riding off in all directions. Too often the review articles are slipshod and uncritical – a succession of index cards blown up into short paragraphs repeating each author's own estimate of his little contribution. Scientific knowledge is public knowledge. It is not enough for the experts to be tacitly of one mind about a fact; this fact must be expressly stated in print and given explicit justification.

My general proposition may also throw some light on the psychology and sociology of scientists and science. There is a good deal of talk about 'the scientific attitude of mind', as if, somehow, scientists were differently made and reared from ordinary mortals. But experience of scientists soon teaches one that they are not, individually, gifted with some special insight into Nature. As everyday persons, in college or office, they are perhaps more realistic, more ready to examine a problem in direct rational terms, than those trained in the literary tradition, but this can sometimes lead them into disastrous insensitivity to the less tangible values of personality and community. Confusion has arisen because we can never really enter the private mind of the scientist at work, following hunches, turning over and over a few uncertain clues, limited by prejudices and misconceptions, wasting months and years after false scents; we judge him only through the public mask that he puts on to write his papers, when all seems good sense, cold logic and brute facts. An honest admission of chance or intuition as his guide would not help to persuade the world that he had the truth; he must paper it over and bolster it up with impersonal objective rationalizations. The apparent suppression of subjectivity in scientific writing is the bed-side manner of the trained professional; it does not represent a lack of pride, prejudice and passion in the man.

The open-mindedness which is supposed to be characteristic of scientists is also an intellectual technique that has to be learned. Experience teaches that it pays to give a fair hearing to new and strange ideas, for these may contain the grains of a new truth. It does not do to become too attached to one's own pet ideas, for a new experiment or a better theory may, with inexorable logic, render them false. Criticism, in modern science, is deliberately conservative and positive, for nothing rebounds more disastrously than an unjustified personal attack or wilful blindness to a serious rational objection. Indeed the wise scientist will already have anticipated, in his own mind, all the critical arguments against his ideas, before he ever publishes his work.

Modern scientists learn these unusual mental habits because they do not work in isolation but have grown up inside a corporate tradition. In my view, Robinson Crusoe could no more make a scientific discovery (as distinct from a valid observation or generalization) than he could deliver a legal judgement. No man can be so detached as to be plaintiff, defendant, judge and Court of Appeal, all in the same cause. This is not to withhold admiration for the genius of individuals who make outstanding contributions. Great scientists are not ordinary men; they have an unfair share of intelligence, persistence, imagination and curiosity. But they have almost always worked within a mental framework built up by their predecessors and contemporaries. The cult of the hero has done much harm in the history of science, and is always something of a simplification. If you want to understand why science works so extraordinarily well as a means of discovering the unsuspected truth about the world, you must look on it as more than a crowd of star performers; it is a highly organized social institution in which knowledge is built up by a process of deliberate intellectual cooperation.

Finally, there are some philosophical implications of my argument. For example, we can offer some sort of solution to the 'demarcation' problem: how to distinguish the utterance of a scientific statement from other intellectual activities. The answer is empirical; it does not depend on some logical criterion – verifiability, falsifiability, objectivity, decidability – but on the context in which it occurs. It is a scientific statement if it is made within a scientific discipline in the furtherance of scientific knowledge. How, then, do we recognize a scientific discipline? My answer is: through the general attitude and mutual relations of its practitioners. They will be counted as scientists in so far as they are willing and able to agree on a set of basic propositions and cooperate to extend the scope of their agreement by rational inquiry and discussion. So there is no objection to talking of the science of history where we are concerned primarily with establishing out of concrete evidence a detailed narrative on which all the experts would agree (however much they might quarrel over the interpretation of events in causal terms). On the other hand, the advertiser who claims that 'scientific tests have proved' the superiority of his product is misusing words unless he is prepared to publish all the results in full and permit independent comment. One would also exclude much of the literature on subjects such as psychology, where there are many quarrelling sects with different doctrines and no central set of facts and theories to which they all subscribe. Again, there is a useful boundary between science, where truth is essentially single-valued, and technology, where there may be a number of equally good ways of achieving a practical end. Measurements of the same physical constant in two different laboratories are expected to agree; but no one expects two different designers to turn out identical aircraft or toast-racks.

This means throwing away much of the philosophical discussion of 'scientific method', conceived as a means for arriving infallibly at the truth. All that we can say about this is that there are a number of linguistic and logical techniques by which a group of people may be freely persuaded into accepting a common view. Logic itself is precisely such a device – a minimal set of agreed rules about the relations between propositions. Mathematics we include, because it mechanizes logic with symbols; classification and definition, so that men shall have a common language; counting and measuring, because they are reproducible by all observers; experiment because it can be repeated by different men; prediction because nothing is more convincing than to foretell the unknown future. But not rhetoric, because not all men have the same strings to their hearts, and not introspection, because we have all lived in different worlds and have different dreams. The philosophers are properly concerned with epistemology – the study of knowledge itself – but it would be a pity if they pretended that the rationale of scientific arguments was different in principle from the ordinary logic of everyday life. It is particularly important to realize that science is fallible. Men are fallible, and it is no more impossible, in principle, for a whole group to be deceived than it is for one frail mortal to make a mistake. The only advantage that science has is that it is like the old Polish parliament: everyone has a veto and is trained to exercise it.

My argument is that science is not an autonomous realm of knowledge, governed by immutable rules, but is a way of doing intellectual business. If I may be allowed an analogy, it is like the Stock Exchange. The price of a particular share on a particular day is not fixed by any one individual, although it may be greatly influenced by big buyers and sellers. The truth value of a scientific theory, also, is not predetermined by the genius of its discoverer but depends on the price which other scientists will put upon it. In science, as on the Stock Exchange, one is bound to be absolutely fair and honest because the long-term profits of decent trading are greater than the immediate gains of fraud. But honesty will not protect any of us, individually, from errors of judgement; in the long run, the market price is the only fair price. There may perhaps be better social means than the Stock Exchange for maintaining the economic health of the nation; I doubt if there is any better instrument for achieving reliable knowledge of the world than a freely cooperating community of scientists.

7

Undoctrinaire Inspections*

The principle of 'consensibility' throws light on the philosophical problem of 'intuition' in science.

Writings about science remind one of the old joke about the international essay competition on the subject of elephants. The Englishman wrote 'Elephants I Have Shot'; the American wrote 'Bigger and Better Elephants'; the Frenchman wrote 'L'Eléphante et Ses Amours'; the Pole wrote 'The Elephant and the Polish Question.' The experimental physicists say 'Science is measurement'; the theoretical physicists tell us that the aim of science is to reduce the universe to mathematics; 'It's all done by mirrors,' complain the logicians; 'Don't stop me; buy one!' is the theme of the technologists; the psychiatric interpretation seems to be 'Look what a lovely mess I've made!'; and some of our more doctrinaire sociologists embroider the slogan 'Give him the money, Barney!' Science is so elephantine that one tends to see it only from the standpoint of one's own particular experience and interest.

The present authors both speak, with immense authority, out of the central regions of biology. Their writings refute the assertion that all scientists are unlettered and lacking in true culture, for they are witty, charming and full of humanity.

Induction and Intuition in Scientific Thought is (as the author admits) a rather long title for a rather short book – no more than three lectures in a semiphilosophical vein. The theme is simple, and not unfamiliar: the traditional description of scientific methodology as the apotheosis of inductive reasoning is shown to be inadequate; the role of intuition in the practice of science is too important for it to be neglected in philosophical principle. Yes, the theme is familiar enough by now (may I suggest Polanyi, Hanson, Körner and Kuhn as recent candidates for an enlarged bibliography of the subject?), but it can bear repetition in this clear,

* From a review of *Induction and Intuition in Scientific Thought* by P. B. Medawar (Methuen, 1969), and *The Relations between the Sciences* by C. F. A. Pantin (Cambridge University Press, 1968), published in *Science,* 166, 729–30 (1969).

incisive, and elegant formulation. Unphilosophical scientists quite rightly turn up their noses at boiled Bacon, which is what many of them still imagine to be the view of science taken by the philosophers; books like this are admirable antidotes to that stale, unpalatable stuff. The argumentation is too tight to be summarized here, but the spirit of it can be judged from the sentence in the final paragraph:

> The scientific method is a potentiation of common sense exercised with a specially firm determination not to persist in error if any exertion of hand or mind can deliver us from it.

Yes! Yes, indeed!

I knew Carl Pantin slightly, as a sweet-natured, courteous, thoughtful and modest man, and it is a pleasure to renew the acquaintance – too late, alas – by reading his book. *The Relations between the Sciences* is an attempt to explain the general 'philosophical' problems that arise in what Pantin calls the 'unrestricted' sciences, such as biology and geology. The treatment is tentative, undogmatic, and illustrated by many practical examples from his own experience. He does not cut as sharply as Medawar, but perhaps his book is the more valuable from the evidence it gives of honest perplexity in the effort to deal with genuine difficulties. One can almost regard it as a piece of natural history in its own right – a record of a journey taken into unfamiliar realms, with descriptions of the peculiar creatures observed on the way.

For example, Pantin too is deeply concerned about the contradiction between the official deductive scheme of scientific proof and the obvious fact that research is guided largely by intuition. Like Medawar, he takes Whewell as his hero. He talks of the ease with which he was able to say 'That is *Rhynchodemus bilineatus*' upon sight of a tiny worm, and contrasts this 'illative sense' with the step-by-step verification of an identification by means of a taxonomic key.

This aspect of scientific methodology is peculiarly obtrusive to the working biologist; but undoctrinaire inspection reveals the same problems in chemistry, physics – even mathematics. Nor are true philosophers especially abashed by being told that more can be learned or discovered about the world than can actually be proven. As I have hinted, this is old hat.

The puzzle, which Medawar and Pantin do not pose, and to which neither hint at a solution, is why anyone should think otherwise. Why are so many ordinary scientists, to whom these books are really addressed, so susceptible to the spell of naïve Baconism that it must be exorcized, with bell, book, and candle, generation after generation? The reason seems to have something to do with a Platonic ideal of 'objectivity'. In these pages, research is described as a purely personal activity, by which a pre-existing Truth is discovered. The findings are not really meant for a human reader, but for the sweet Princess Reason herself, in her high castle, to say yea or nay.

This is an unattainable standard of proof. The fact is that science is a purely human pursuit, and the best we can achieve is systematic intersubjectivity or 'consensibility' – an agreement that 'thus it is' between well-informed minds. Logical deduction from hypothetical premises – the scheme which Medawar favours, following Popper – is then only one of the forms that the persuasive communication of knowledge can assume. The human brain has an unequalled capacity for the recognition of meaningful patterns, and there can be certain and complete agreement between two persons concerning the congruence of patterns that they both perceive. Surely the illative sense is no more than the introspective occurrence of this mental phenomenon – a comparison of such potentially consensible patterns within the memory of the same individual? Or must we pay tribute also to Princess Rhyme, who dwells on the same distant crag?

For the neo-Baconian, however, the controversy now centres about the logical status of the recognition of congruent patterns. He would, one supposes, like to reduce this activity to a succession of simple operations, along the lines of a mathematical formula. He wants to apply a Turing machine – an idealized logical computer – to the problem. But we know now, from bitter experience in attempting to construct actual programs for pattern recognition with real computers, that this is an extraordinarily difficult task, of which we can say only that it ought to be possible in principle. Until a practical procedure of this kind has been found, we do better to rely upon the undoubted fact, emphasized by Pantin, that living brains – animal as well as human – have this power to a very high degree, and that scientific knowledge, acquired, verified and critically assessed with its aid, is as certain as is humanly possible.

We approach here, indeed, another of Pantin's problems – the status of emergent characteristics of very complex systems. Because the biologist fastens his attention on such systems – cells, organisms, societies of individuals – he must face up to the present impossibility of cerebrating the phenomena of life and mind from the known properties of the elementary constituents of such systems. As Pantin points out, there exist complex inanimate systems, such as thunderstorms, which show many of the features we associate with living things, but these lack the fantastic functional organization, part by part, limb by limb, cell by cell, organelle by organelle, down to the molecular level, to be found in all living creatures. Here he wisely refrains from dogmatism, whether mechanistic or vitalistic; the point is to observe how scientific progress can uncover relations of structure and function at all levels, and thus gradually change the terms in which these problems are posed. The discovery of the structure of DNA, for example, does not solve the puzzle of heredity, but at least we can now avoid such mystical concepts as 'germ plasm' and 'gene' in our attempt at a description.

There is much else of value that cannot be summarized here. In the interdisciplinary essay competition proposed at the outset of this review, the biologist often does no more than croon 'Ripeness is All!' The best recommendation I can give to the present books is that they are different. They present science through the eye of the biologist, neither blinkered nor sentimentalized. They have the power to provoke further constructive thought in the mind of any well-educated reader.

8

Is Science to be Believed?*

The credibility of scientific knowledge does not depend on any absolute logical principle, but upon the coherent, consensible world picture that it presents.

The disciples of Immanuel Velikovsky claim that the science of astronomy is all wrong.[1] From their writings one might gather the impression that science has become dogmatic, and that the truth is being deliberately suppressed by action of the scientific authorities. They talk the language of victims of witch-hunting, censorship, etc.

I am not concerned for the moment with Velikovsky's actual views on the motion of the planets. The question we should ask is: 'What is the ultimate justification for a fairly firm belief in scientific knowledge?'

Strictly speaking, this is a philosophical question, in the technical field of epistemology. The question has certainly been studied at great length, by many people, for many years; yet the conventional philosophy of science has still not given us a reliable answer. Many of the old arguments, such as those based on the principle of 'induction' (for example, the sun will surely rise tomorrow, because it has risen 10 000 000 times before, etc.) are logically incomplete.[2] Every attempt at a proof falls back on intuitive axioms which might themselves be false.

I prefer, therefore, to start from the characterization of science as a social product.[3] Men cooperate to construct a great body of knowledge which they hold in common. Scientific knowledge is *public* knowledge; it is what all men can agree to be the case. The work of scientists is the creation, the criticism, or the correction of a consensus.

To show the power of this idea, let us consider some typical features of scientific activity.

Mathematics

It is often said that 'real' science is essentially mathematical. This, as will be demonstrated later, is not true, but mathematics certainly plays a very

* The substance of a Thursday Evening Discourse, 3 May, 1973, at the Royal Institution, published in the *Proceedings of the Royal Institution*. This lecture eventually expanded into *Reliable Knowledge* (Cambridge University Press, 1979).

important part in science. This is because it is a highly 'consensual' language. Every symbol and operation is precisely defined, and has the same meaning for all men: mathematical reasoning is the lowest common factor for all communication. Remember that the proof of the theorem of Pythagoras is as clear to us now as it was to the Greeks, 2000 years ago, and that it was discovered independently by the Chinese. In the seventeenth century, science was communicated in two universal languages – Latin and algebra. Nowadays, Latin is still used for formal botanical descriptions, but has been replaced as a general language of scientific communication by Broken English.

Experiment

An experiment is not merely a new, contrived experience; it is also a well-defined and repeatable event, and hence a suitable component of a body of public knowledge. The event can be shared because the circumstances have been so closely defined. If the result is surprising, this reproducibility has great persuasive power: you can always go and do it again yourself, and see if you get the same answer.

The experiment by which Rutherford split the atom can be repeated in any high school physics laboratory, with home-made apparatus. The same cannot be said about many modern scientific experiments. In high energy physics, each experiment may cost about £1 million a time, so we must rely very heavily on the precision of the apparatus and on the integrity of the research workers who carry it out. Typically, this is the work of a *team* of scientists, another sign of the social character of science.

Repetition, verification and the checking of results are very important in the validation of scientific knowledge. Good experimental data are not just the product of the lone research worker: they are really social products, generated by the cooperation of many independent minds. The role of *criticism* in science is often undervalued by philosophers and popularizers.

In practice, even the results of very big, expensive experiments are not always clear cut. When statistical uncertainties are present, as in high energy scattering experiments, sophisticated interpretation is necessary. Thus, the question whether there was a dip in the peak of the data for the A-2 meson was, for quite a time, undecided, despite the various expensive experiments carried out to confirm or falsify the original observation.[4] Problems of statistical interpretation are especially acute in fields such as psychology, sociology and medicine, where it is very difficult to design an experiment that will really settle some doubtful point. Anyone who has looked at the evidence put forward by Ernest Sternglass,[5] in support of his conjecture that nuclear tests were having a big effect on infant mortality, will know what is meant. Do the peaks and kinks in the gross infant mortality curve really have any connection with the starting and stopping of nuclear tests?

Prediction

Ideally, scientific knowledge is a complex pattern of experimental facts explained or illuminated by a mathematical model. What makes such explanations convincing? Obviously, general agreement between theory and experiment is very satisfying, but one must always be conscious of the possibility that the theory may have been tailored to fit the known facts. Professional theoretical physicists are trained to look for adjustable parameters in other people's theories that might have been used for this purpose.

The most convincing argument for a theory is a successful prediction. Thus, when Gell-Mann and Ne'eman put forward classification schemes for elementary particles that implied the existence of the omega-minus particle, they could not have known that this particle would actually be discovered experimentally two years later.[6] Logically, this is not very different from confirmation, or 'retrodiction', but psychologically it is much more compelling.

But note the danger of the *false* confirmation of a prediction. The theoretical physicists have very good reason for believing in the existence of *quarks* – heavy particles of fractional charge. A group of Australian cosmic ray observers obligingly produced some cloud chamber tracks that seemed to have the required properties, but this evidence is now considered unreliable.[7]

Discovery

This word is often used as a synonym for 'research'. But what does it mean? In my opinion, something quite complicated, such as 'the falsification of an assumed null theoretical prediction'. Columbus falsified the prediction, made on no sound basis, that there was nothing but ocean to the West, at least until one reached China. The coelacanth was surprising because it was assumed to be extinct. Discoveries are, of course, the most exciting events in research; they can also be very powerful and persuasive. The best science satisfies Beethoven's dictum 'in music everything should be both surprising and expected'.

But again, not all 'discoveries' are sound. They need very critical testing before they become part of the consensus of science. A recent example is the phenomenon of 'anomalous water'. A distinguished Russian physical chemist, Deryagin, reported that water condensed from the vapour in a fine glass capillary had unusual properties, such as high viscosity, strange melting behaviour, different Raman spectrum, etc. This observation was reported quite sincerely, and was confirmed by further experiments of the same kind, in various countries.[8] Specific theories were then put forward to explain the unusual properties – for example, that the water was catalysed, by contact with the glass, to a special polymeric structure. The name 'polywater' was given to the material, and it was proclaimed as a very significant discovery, an ideal topic for research, etc. But sceptics remained. Joel Hildebrand said he found it 'hard to swallow'! There were more experiments, which finally showed that the material almost certainly contained a high concentration of impurities. Thus, the infra-red spectrum of one specimen was almost identical with the spectrum of sweat, suggesting that 'polywater' was, like genius, '10 per cent inspiration; 90 per cent perspiration'. I am unsure whether the nature of the dissolved silicates or dirt is yet fully understood, but that would be a minor scientific problem.

This story is a warning against believing all the 'results' of scientific research. We must always remain sceptical and critical, ready to repeat the experiments or observations many times, in many different ways, until the results are fully confirmed. The body of scientific knowledge is not just the sum of all pointer readings and conjectures; it means those facts and principles that have been thoroughly tested, criticized, verified and unfalsified, and that are very widely accepted by a large community of independent experts.

Observation

The conventional philosophy of science can deal more or less satisfactorily with the above examples because they lie in the realm of physics and mathematics. Consider, however, a descriptive science such as biological taxonomy. There is no difficulty in recognizing plants from their drawings in a herbal or flora. In fact, we can turn to a completely foreign collection of botanical drawings, such as a Chinese pharmacopeia,[9] and still identify the plants represented in it. In other words, such representations are *consensible*: all men have the innate ability to recognize patterns which they have seen before, or to agree that two such patterns are essentially equivalent. Knowledge gained in this way is perfectly 'scientific', according to my general definition.

But remember that pattern recognition cannot be reduced to a mathematical formula. Great difficulties are being encountered in the attempt to perform this perfectly ordinary human act with a computer. There is no simple logical algorithm by which one can 'prove' that two patterns – a plant and its drawing in a flora – are the same. This faculty thus tends to be ignored by the philosophers, who always want to reduce everything to formal logic. This seems to be the so-called 'intuitive' element in science, which is often stressed by biologists but 'pooh-poohed' by the physicists.

Yet where would chemistry be without visual patterns? For example, try to describe in words any complex chemical compound and its attendant reactions; the task is practically impossible without drawing a picture of the formula in two – or even three – dimensions. Even physics is heavily dependent on descriptive geometry. The most accomplished of mathematical physicists would find it impossible to set up an algebraic interpretation of, say, the patterns of disclinations in a cholesteric liquid crystal; we must have pictures to *think* about such a system. The extraordinary power of the scanning electron microscope, whose output looks like a simple three-dimensional perspective photograph of solid objects, depends on the same natural facility which all men share.

This point was recognized long ago by Leonardo da Vinci who wrote:

> And ye who wish to represent by words the form of man and all aspects of his membrification, get away from that idea. For the more minutely you describe, the more you will confuse the mind of the reader, and the more you will prevent him from a knowledge of the thing described. And so it is necessary to draw and describe.

That is what he did. There is a fantastic difference between Leonardo's anatomical drawings and those current in his day, which were based on Galen's descriptions and on occasional dissections. Leonardo saw, and drew, what was really there. He was the finest scientific observer that ever lived. It is a tragic irony that his anatomical work was not published until centuries later, so that it made no contribution to science after all.

Reality

This takes us to a deeper point. To the layman, scientific knowledge seems esoteric and strange; it has almost the power of magic. It is difficult to believe that it is not inspired from some hidden source, only to be acquired by mysterious rites based upon transcendental doctrines. This is not at all the case. Even a sophisticated, highly theoretical subject such as quantum physics really depends on vast quantities of very direct material observation and experience.

The scientific world picture is merely an extension of the everyday world, encompassing the very small, the very large, the very rapid and the very slow. If one turns to works on child psychology – especially the books of Jean Piaget[10] – one finds a most interesting discussion of the way that a child acquires concepts of number, size, space, time, etc. Two factors are at work:

(*a*) personal experience and manipulation of objects, and

(*b*) social contact, by means of language, with older people who have already learnt to give meaning to such experience.

The result is a gradual growth of the ability mentally to structure the world about us, both in geometrical patterns and in abstract words. The reality of the child's home extends, by experience, to the town in which he lives, the country to which he belongs, and, eventually, to the world at large.

Scientific knowledge, for each individual, is essentially an extension of this making of maps. The physics student, for example, starts with magnets and iron filings, and learns to think in terms of 'fields of force'. The anatomy student sees various lumps of flesh, and learns to recognize them as diseased or healthy organs. In each case, he acquires this new way of thinking both by personal experience and manipulation and by discourse with those familiar with the subject. When you have lived in it for a while, the world of science seems just as matter of fact and real as your own kitchen.

Error

But science is no more infallible than ordinary daily observation. In fact, because of the very boldness of its ventures, and an occasional failure to curb speculation, it can get things quite wrong. The official histories of science emphasize its self-correcting power, especially when fundamental inconsistencies come to light, as in Lavoisier's attack on phlogiston, Rumford's experimental refutation of the caloric theory of heat, the Michelson–Morley experiment and the theory of black-body radiation. From such inconsistencies new syntheses arise.

The assumption is often made that it is sufficient to point out the contradiction to initiate the process of correction – but this can be a slow process. For example, an implicit principle of geology was the permanence of the main continental pattern.[11] Why should one assume otherwise? For this reason, the obvious geographical fit of the continental margins, which had been known for centuries, was dismissed as a mere coincidence. Wegener's theory of continental drift, first put forward around 1910, seemed against the pattern of agreed facts, and was therefore treated with scorn.

Notice how such an attitude was reinforced by the social character of science. Knowledge on a subject such as this cannot easily be experienced personally by the individual student, and is therefore learnt as a 'fact' in the course of a scientific education. The role of authority cannot be neglected: in the appointment of a lecturer to a university post, for example, there would be some emphasis on his holding 'sound' opinions on such topics. A zoological colleague, Professor Howard Hinton, had such an experience when a paper that he submitted in 1938, discussing the zoological evidence for continental drift, was rejected for this reason. It is interesting to note that the geologists from the southern hemisphere, with closer personal experience of the most direct evidence, were much earlier to accept the drift hypothesis.

In fact, the major argument used against the theory was the mathematical theory of the dynamics of the Earth. Harold Jeffreys showed, on physical grounds, that the suggested mechanisms were not sufficiently powerful – therefore, it would seem, the phenomenon could not have happened. This sort of argument was particularly convincing to geologists because it was couched in algebraic symbolism which they probably did not understand. They evidently had much less faith in their own type of 'intuitive' evidence, based on pattern recognition, such as the distribution of various animal and plant species or the continuity of geological formations.

Evidence and paradigms

How could so many intelligent, well-educated people deceive themselves for so many years? The interpretation of evidence depends on one's training, on motivation, on the context and on the point of view. We are all familiar with mysterious-looking photographs, whose meanings only become clear when we are told that they are unconventional views of perfectly normal objects. We are familiar with optical illusions that make two equal figures seem quite different in size – for example, against a background that suggests strong perspective. By the same tokens, it is very difficult to set aside the background experience of a scientific education, where many irrelevant facts have been carefully classified according to some conventional theory, and look at a question 'objectively'.

Another metaphor (first pointed out some years ago by N. R. Hanson,[12] whose death at an early age robbed the philosophy of science of one of its most attractive and imaginative personalities) is the ambiguous drawing, the face that seems sometimes to be that of a young woman, and sometimes of an old hag. This exemplifies the concept of *paradigm switching,* enunciated by T. S. Kuhn.[13] Our understanding of scientific evidence must switch over from one pattern of interpretation to another, discontinuously. Without significant clues, and a conscious effort, this can be almost impossible; in fact, one sometimes cannot see the alternative at all, even when told. Yet, when the switch occurs we do not destroy past knowledge; we merely restructure the pattern, and change the meaning and emphasis of the various features.

Let us return now to Velikovsky. His main theme is that the astronomers are ignoring significant evidence from historical and prehistorical sources, such as the Bible and the Chinese Classics. On the basis of this evidence, he has put forward an alternative picture of the behaviour of the planets, and claims new explanations of phenomena, including some general predictions that have been verified.

Yet the scientists completely reject this interpretation. From the point of view of physics, it looks completely nonsensical because it is quite inconsistent with the world picture that has been built up over the centuries. Most seriously, it completely ignores the numerical, quantitative analysis of the motions of the planets, on which, for example, was based the practical achievement of sending a rocket to the Moon. To accept the Velikovsky cosmology implies the rejection of most of current physics – not a step to be taken lightly. This is evidently a greater paradigm switch than can be tolerated or achieved – and we are not bound to accept an argument just because it is new and exciting.

But the attempt by some astronomers to prevent the publication of Velikovsky's books was misguided. Science is to be believed *because* it can be openly criticized. The criterion for belief is not 'authority' nor 'logical necessity', but 'nobody has yet found a more convincing pattern of interpretation'. The patterns that scientists currently accept may not always fit together without contradiction. Each little region may be well understood, yet consistency may be lacking over a wider field. There are undoubtedly numerous errors and fallacies to discover and plenty of rude shocks for our complacent beliefs to suffer. That is why science is worth studying. But, for all its defects, science is at least as credible as chalk and cheese: what with substitutes, additives and preservatives, even these can scarcely be trusted nowadays!

References

1. Velikovsky, I. (1950) *Worlds in Collision,* Doubleday, New York.
2. Körner, S. (1966) *Experience & Theory,* Routledge & Kegan Paul, London (for example).
3. Ziman, J. M. (1968) *Public Knowledge,* Cambridge University Press.
4. Schubelin, P. (1970) The puzzle of the A-2 Meson, *Physics Today,* November, pp. 32–8.
5. Sternglass, E. (1972) *Low-Level Radiation,* Ballantine Books, New York.
6. Discovery of Ω^-, *Physical Review Letters,* February 24, 1964.
7. Jones, L. J. Search for Quarks, *Physics Today,* May 1973, pp. 30–7.
8. Anomalous water, *Physics Today,* October 1970, pp. 17–20.
9. Needham, Joseph (1965) *Science & Civilisation in China: Vol.* 1, Cambridge University Press.
10. Piaget, J. (1929) *The Child's Conception of the World,* Routledge & Kegan Paul, London.
11. Hallam, A. (1973) *A Revolution in the Earth Sciences,* Clarendon Press, Oxford.
12. Hanson, N. R. (1958) *Patterns of Discovery,* Cambridge University Press.
13. Kuhn, T. S. (1962) *The Structure of Scientific Revolutions,* University of Chicago Press.

9

Mathematical Models and Physical Toys*

Research in theoretical physics demands a subtle combination of mathematical analysis and physical intuition.

Philosophers, so it is said, spend much time and effort discussing the relationship between analytic and synthetic propositions – or, as we might put it more crudely, between theory and experiment. Most scientists, fortunately for their peace of mind, do not worry too much about such questions, or if they do they satisfy themselves with some handy piece of conventional wisdom and go on their way. But, as a so-called 'theoretical' physicist (although I should much prefer to be called simply a mathematical physicist, without any suggestion that theories are not the business of those who make observations and experiments), one cannot remain quite indifferent to this problem in one particular aspect – the relationship between mathematics and physics. For us this is not an abstract philosophical question; it is our daily task to put numbers into Nature – or to find them there, or to invent them, or to put Nature through our mathematical sieve, or cut it down to fit our Procrustean bed, or whatever it is we are doing. I wish to discuss how this is done in practice, and to make some comments, as a professional player, on the rules of the game.

I shall try to make the examples on which I build my argument as simple as possible; if I fail, I trust that the abominable sophistication of the subject, which it takes so many weary years to learn, will be blamed, rather than that I, and my colleagues with me, be accused of deliberately speaking above the heads of ordinary people. It would be nice to be a real philosopher; then I could use everyday objects, like tables, or spoons, to illustrate my argument; unfortunately, even so simple an object as a table is not easily described mathematically; so we have to talk about electrons, which we know more about.

The tasks of the theoretical physicist are set by the experimenters. 'Tell us,' they ask, 'why we got this value for our measurement' or 'What answer shall we get if we do such-and-such an experiment?' Sometimes,

* Inaugural lecture as Professor of Theoretical Physics at the University of Bristol, delivered on 28 January, 1965, and published in *Nature,* 206, 1187–92 (1965).

of course, we try to tell the experimenter what he should do before he even asks us, and if it comes out right they give us a Nobel Prize or something; but that is unusual – and it does not really alter the job that we have to do.

But before we sharpen our pencil, hang our jacket behind the door, and sit down quietly before a block of clean white paper, we ourselves must ask a few questions. For example, we might ask the experimenters what they thought they had measured. What was the apparatus made of? Did it work well? Was the sample pure? Was the magnetic field uniform? – and so on. This inquiry, so often taken for granted and so often unwisely neglected, is needed because the laws of physics are only clearly stated for ideal situations – situations qualified by phrases such as 'in the absence of other forces' or 'for a perfect crystal' or 'assuming that all processes are reversible'. A good experimenter knows these rules, and tries to get as near the ideal as possible.

One must truly admire the extraordinary skill of the modern experimental physicist. He can detect neutrinos, which weigh almost nothing and are so difficult to catch that they can travel through the thickness of the whole Earth without noticing it; he can measure some quantities to an accuracy of one part in ten thousand million, which is less than an inch in a distance from the Earth to the Moon. His command of technique is astounding. If we train him properly he should have sufficient grasp of the theory to have anticipated most of our objections. One can only regret the snobbery that makes the experimental physicist a sort of handmaiden to the theoretician; that idea ought to have gone out with the Greeks. The experienced observer of Nature often perceives what is going on in some new phenomenon, and gives an interpretation – a 'theory' – long before the arm-chair scientist can dream up a mathematical analysis to bolster the explanation.

Yet even when the experimenter has done his best, we may still be entitled to remark: 'Your system is very complex. What can we safely ignore?' We may have to enter into a long discussion trying to judge whether the deviations from the ideal really matter, before we even begin our calculation.

For example, suppose that I am asked to calculate the boiling-point of water. Can I assume that it consists only of molecules of H_2O? The experimenter tells me that his samples do, indeed, contain small traces of various impurities, but that he has found that changing the amount of impurity does not make much difference. Of course, it does make some difference, and he stores that up for another question to be answered one day. Anyway, we decide that this is irrelevant.

Now at this stage, in many fields of physics, we may try to answer the tiresome experimenter by saying, blandly: 'The following is the physical set-up of the problem. It is governed by such-and-such laws of Nature. The answer you want is therefore just the solution of the following equations.'

For example: 'Your system consists of about a million million million million molecules of H_2O, each of which is shaped rather like a boomerang. The molecules attract one another with forces varying with distance, as follows' (here show a lantern slide of a complicated graph). 'We may assume the laws of classical statistical mechanics. Therefore, all you need to do this is to minimize the free energy, which is just the following formula' (here you write it on the blackboard, but take care to erase it before they ask what it means), 'and Bob's your uncle.'

This answer is two things, one of which is practically useless. It is, first of all, a solution 'in principle'. That is, it tells us that a man, or a machine, who could do mathematical computations a million million million million times would eventually get the number we are looking for. Perhaps this is true – but it is not a proposition that can readily be decided. In fact, the only way I can imagine of deciding it is to make an analogous system, obeying the same laws, and then to see how it behaves – put another pint of water in another saucepan, and bring it to the boil.

But this 'solution', however impracticable, is also an essential step, for it constitutes a mathematical formulation of the problem. We have brought our system, our pint of water, out of the intuitive realm, and under the rule of mathematics. It has been abstracted and symbolized. It is no longer an assembly of empirical observations, but an object, or a subject, for theory.

My guess is that this is the step which interests the philosophers. But for the practising mathematical physicist it is not the most difficult step in the argument; it is not the point where his particular skill or artistry shows itself best. The real problem for him is to reduce these equations, so multifarious and complex, with so many variables all linked with one another, to a form in which they can actually be solved.

It is well to remember that the range of mathematically soluble equations is extremely limited. Even with the aid of a powerful computer one can only handle three or four variables at once. The number of operations that need to be performed for an exact solution of a set of equations in, say, a dozen variables is enormous. Suppose, for example, that we give only ten different values to each variable – much too coarse a grid for any sort of accuracy – then we might have a million million possible sets of values to investigate, which would take a million seconds if we looked at each one only for a microsecond – and that is about 300 hours of machine time. It would take a very large and expensive project – say, the design of an aircraft, or a nuclear power station – to justify such expenditure. Yet this is the complexity that our mathematical formulation would imply if we were in the process of solving exactly the problem of just five objects moving under each other's mutual attraction in space and we were told to sum some mathematical quantity over all their possible configurations.

So we must make approximations – drastic approximations. We must somehow reduce our innumerable equations, with all their interacting variables, to a few simple ones, the solution of which we can compute; and we must do this in such a way, with such care and skill, as not to lose the answer to our problem – for remember that we are eventually wanting to say that the boiling-point of water, under ordinary conditions, is about 100 °C. Not only do we want a good answer, but we want some assurance that it is a good answer, that we have not made serious errors in our calculation, just as we insist that the experimenter should make his measurements accurately.

There seem to be two different ways of constructing approximations in applied mathematics and theoretical physics. The first method leans on pure mathematics, where we often represent the answer to a problem by a convergent infinite series. That is to say, we construct the solution as the sum of an infinite sequence of separate terms, each of which we can calculate exactly; the terms get smaller and smaller as we go on, and we can get an answer as close as we like to the true answer by working out a sufficient number of terms. If, for practical reasons, we have to stop after, say, the twentieth term, then we have got an approximate solution, as we should say, 'to the twentieth order'.

What we do, therefore, is to look at the equations of our mathematical formulation, and try to see whether they can be cast into a form that will yield such a series solution. In particular, we may look at all the equations and try to decide which terms and equations are least important. We drop these for the moment, and see if the remaining terms and remaining equations are simple enough to be solved exactly. Then we see what the effect would be of adding in the next largest terms, and so on.

An obvious example of such a procedure is provided by the most famous problem of theoretical physics – the motion of the planets about the Sun.

In principle this is a set of equations for many more than five bodies, all moving under their mutual gravitational attractions; yet it was solved very exactly indeed, long before the days of computers, with the help of tables of common logarithms. The key to the solution was, of course, that much the largest terms in the equations are those corresponding to the force exacted by the Sun on each planet. If we drop all the other terms, we are left with some very simple equations; each planet just goes round the Sun in an elliptical orbit. Then we can treat the interplanetary forces as causing small 'perturbations' which slightly change the shape and orientation of each orbit, but which can be calculated quite accurately. From these new orbits we can recalculate the perturbations more exactly – and so on.

Such a mathematical approximation is valuable because its accuracy can often be checked internally. Thus, we may be able to show that the procedure we have used to get a fair approximation may be repeated again and again, to give a sequence of closer approximations to the true answer. This is equivalent to constructing a series solution to the problem, and hence we can use some of the methods of pure mathematics to estimate the margin of error that we might have made, or to show that our procedure really could, as we say, 'converge' on the true answer. In principle, we should not then have made any approximation at all; it is a rule of pure mathematics, hallowed by centuries of usage, and justified by the mysteries known to undergraduates as 'epsilontics', that a convergent series is as good as a simple finite algebraic expression for a solution – which is just as well, because almost every expression that appears in any formula in physics, such as a cosine, or a logarithm, is itself only definable algebraically as the sum of an infinite series. In practice, of course, the mere existence of a convergent series solution is not enough; it must converge rapidly enough so that we can get a good idea of the total from the sum of the first few terms.

If pure mathematics alone may not always give a clue to a good approximate solution, we may sometimes discover such clues in the physical phenomena themselves. An example suggests itself in meteorology, where a complete solution to the motion of large masses of air subject to dynamical and thermodynamical forces is of great practical importance. This is a problem that looks very intractable if we simply write down all the equations and try to cerebrate a scheme of approximations. But observation has shown that certain structures – 'depressions' and 'anticyclones' – tend to persist, and move more or less as permanent entities. These structures are, of course, gigantic whirlpools or 'vortices' in the atmosphere, and are characterized, mathematically, by a quantity called the 'circulation' or 'vorticity'. This suggests that the equations of motion should be cast into a form in which this vorticity appears explicitly – and it turns out that a scheme of approximations can be constructed with this circulation equation as a starting point.

It is interesting to note that if we had been looking for a different physical phenomenon, such as sound waves in the atmosphere, we could have produced quite a different scheme of approximations for the same original equation and found just what we had been looking for. Of course, this phenomenon occurs on a different scale, and under different conditions, from the phenomena of large-scale turbulence that we call the weather – but both types of behaviour are implicit in the same equations. There is an extraordinary richness concealed in some mathematical formulae – a richness that often surpasses our power of analysis and prediction. Only too often the experimenters discover whole new régimes of behaviour that were simply not suspected by the applied mathematicians. The fascination of the physics of plasmas, which are only rarefied gases made conducting and carrying electric currents, is that they make us look again and again at a set of equations that any undergraduate physicist can easily understand, and make us see the marvellous complexity and contrariness of behaviour that they can represent.

Nevertheless, this procedure of looking for mathematical approximations is only one of the ways of finding a workable solution. The other way, more typically that of the theoretical physicist rather than of the applied mathematician, is to look for what I shall call a physical approximation. By this I mean that one carries further the process of idealization that has already been used in producing the mathematical formulation of the problem. That is, we deliberately drop some features of the set-up – features that might well be important, if only we could see our way completely through the problem – in order to simplify the equations. For example, in our original problem of trying to calculate the boiling-point of water we might drop the boomerang shape of the molecules, with all the geometrical complications that this leads to, and simply treat each molecule as a hard sphere.

I propose to call such a simplified system a 'mathematical model', and to distinguish it, in principle, from the 'mathematical formulation' which we were previously considering. The essential property of a mathematical model is that it should be sufficiently simple for an exact solution to be found for the equations describing its behaviour. We do not guarantee that this solution is an accurate prediction of the behaviour of the original physical system that it has designed to represent; we may have had to throw away too much cargo when we lightened the ship to get it over the rocks. We may not even be able to show that the equations for the model are approximations to the equations that we wrote down in our original mathematical formulation, for it may be extremely difficult to estimate the errors that have arisen in the process of idealization.

This sounds very unsatisfactory, and vague, yet the real art of the theoretical physicist may lie in his skill at devising models that are mathematically soluble and that still contain the key to the behaviour of the system that he is studying. The justification of a model is not that it is an approximate solution of the equations – a solution that is internally self-consistent, convergent, etc. – but that it provides qualitative or semi-quantitative results that are in agreement with observation. It provides explanations and insights; it can tell us what is going on, even though it cannot be anything like as complicated as events in the real world.

To return again to our example – the boiling of water. The fact is that this is a singularly difficult problem of theoretical physics, however familiar it is in everyday life. The interesting question is not really to calculate the precise temperature at which this occurs – for suppose we have done this for water, then why not for alcohol, or benzene, or mercury, or any of an endless list of different liquids? The problem is to understand why this happens so sharply at a definite temperature – why, for example, the liquid does not just expand and expand, as it is made hotter, until it has become a gas. If we could find a solution to the equations for the very simplest liquid, such as our assembly of hard spheres, that showed this striking mathematical property, then we should be well on the way. But even that still eludes us. All our models either expand continuously, and do not really boil, or they are so abstract, and unlike a liquid at all (for example, they may have no mechanism for flowing freely), that we do not feel convinced that they have any relevance to the problem.

Recognition of the logical status of models is very important. There seem to be two major patterns to which good work in theoretical physics must conform. It must either be a valid mathematical approximation to a calculation of the properties of the system, derived from the mathematical formulation, or it must be an exact solution for the properties of a model. If we choose to work from the original equations of the formulation, we must not, arbitrarily, without mathematical justification, throw away inconvenient terms, but we need not explore all the deeps of possible behaviour before settling on a scheme of approximation that seems appropriate to the phenomenon that we are trying to analyse. For example, in our investigation of the motion of the planets, we need not seriously consider what would happen if the orbit of the Earth passed quite close to Jupiter, where it might be captured as a satellite; we can use a scheme in which the attraction of the Sun is always dominant because that is what happens in Nature. (Of course, we are entitled to examine other cases if we like, as little exercises, or as examination questions, just to test our mathematical skill, but we are then escaping from the realm of physics as a science.)

On the other hand, if we use a model that is already rather divorced from reality, we shall have to do our mathematics very carefully, to show that it does have precisely the qualitative features of the system that it represents. In particular, we must be very careful that we do not construct some closed system of logic, in which the model has been defined at the start to have just the properties that we are going to deduce; and we must also be careful not to try to justify an approximate solution of the equations for the model by an appeal, again, to the phenomenon which we are looking for. This sounds rather stupid – but it often turns out that this is just what lots of people do when they work away, unthinkingly, at some system of models in theoretical physics. I could give some examples of this, in the context of our question about the boiling of water – but perhaps it might become too technical, and again might not do justice to the efforts of many very intelligent people to solve a most intractable problem.

Consider some of the models in which theoretical physics abounds, so as to make the argument more concrete. For example, there is the idea of a gas as an assembly of point particles – that is, of molecules having mass, velocity, energy, momentum, and position but no significant size – which was the clue to our understanding of the laws of behaviour of gases, such as Boyle's Law. It is interesting that nearly 200 years passed between the discovery of this empirical law and its interpretation – yet the mathematics of the atomic model are trivial, and would certainly have been easily intelligible to Newton and his contemporaries. Making good models is not easy; it takes insight.

A more recent subject for model-making is the atomic nucleus, which can be thought of as a little cloud of particles – neutrons and protons – moving under each other's mutual attraction. In fact there are two models of the nucleus. One is called, for technical reasons, the 'shell model'. In this model the particles are thought of as moving nearly independently of each other, as if the forces between them are smeared out into some sort of average field of attraction which keeps them all together. This field is supposed to be strong enough to dominate the motion of each particle, just as the gravitational field of the Sun dominates the motion of the planets – and so we can solve the equations by much the same scheme of approximation as we use for the Solar System. Of course, these are quantum mechanical equations, not the classical equations of Newton, so they do not have quite the same sort of solutions, but the general principle is the same.

The other model of the nucleus is called the 'collective model', or the 'liquid drop model'. In this model we assume that each particle in the nucleus has a well-defined size, and that the particles tend to pack themselves together, like molecules in a liquid, at a nearly uniform density. A nucleus is then to be thought of as a drop of 'nuclear matter', subject to the equations of hydrodynamics. Such a drop is capable of oscillating in various ways, and is also liable to divide in two if it is too big – hence the phenomenon of nuclear fission. Again the mathematics of this model is relatively simple, having been originally developed by the astrophysicists to discuss the stability of stars.

Those two models are interesting because they are mutually inconsistent. In a liquid the forces between neighbouring molecules are the dominant ones – for example, they are the means by which the material resists compression and passes on waves. In a gas, as assumed in the shell model, the particles are nearly independent of one another, and are supposed to collide only occasionally. It is a flat contradiction to suppose that such collisions are infrequent and yet that each particle is very much aware of the pushes and pulls from its immediate neighbours.

Such inconsistencies are typical of really difficult and interesting problems. What we have done, in fact, is to use different models, different physical approximations, to explain different properties. It turns out that some of the properties of nuclei – for example, the stability of certain nuclei containing special 'magic' numbers of particles – are features of the shell model, while others, such as nuclear fission, are explicable from the collective model. We cannot get, say, the magic numbers out of a liquid drop model, nor can we predict fission from the shell model. What we must hope for, of course, is to think of a new model, or a scheme of approximations, that will embrace both models, and give both sets of phenomena at once. They say that such a scheme is making its trial runs – but we must wait and see whether it will come up to its design specifications when it is in regular service.

Let us give another example of mathematical models, right outside physics. In modern economic theory we encounter many attempts to put real mathematics into the system, and to calculate the behaviour of markets, etc. Perhaps the most interesting such attempt was made by von Neumann, a theoretical physicist of towering eminence, who invented the theory of games. He and his successors investigated the mathematical properties of certain types of elementary 'game', with idealized players and rules, and showed that these could show, in a schematic way, some of the behaviour of economic systems. However meagre may have been the results so far (and however misleading may be the work of those who have assiduously applied this technique to the analysis of military situations), this model is valuable. It shows that one might make progress, here as in theoretical physics, by examining the behaviour of very simple, but very precisely defined, systems, quite abstracted from the real world yet still having the one or two essential features which really dominate the phenomena.

It looks much more like real applied mathematics than those dreary attempts to represent economic situations by systems of innumerable differential equations relating unknown functions of imponderable variables. These are not so much mathematics as graphical demonstrations of verbal relations like 'more' or 'less', and they prove very little because one can never feel any confidence in any of the terms which went into the equations in the first place.

What is one to think of another development of mathematical economics – the theory of input–output analysis? Is this a model, or is it a scheme of mathematical approximations? From the outside it looks like a brilliant exploitation of a mathematical property of accounting schemes – that they must balance. A great deal can be deduced from this rule alone, just as many problems of physics can be solved by inspired application of the Law of Conservation of Energy (and much can be deduced about the behaviour of motor traffic from the law of conservation of vehicles!).

Input–output analysis is, perhaps, a model in a different sense, a 'scale model', or a 'working model', or an 'analogue'. One can demonstrate some pretty tricks with gyroscopes, magnets, inclined planes, etc. These devices do indeed have tremendous value, as guides to thought, and sometimes as 'analogue computers' which do the arithmetic for us, without thinking, because they obey just the same equations as those of the system we are investigating. Input–output analysis gives us a small-scale model summarizing all the items which appear in bank accounts, stock inventories, warehouse receipts, and so on throughout the nation. We do experiments on it, and test its properties, just as we might test the properties of a new ship by experimenting on a model in the big tank at the National Physical Laboratory.

A mathematical model, as I conceive it here, is rather more abstract than this. It is a mathematical construction obeying its own idealized laws, not the rough-and-tumble laws of actual circumstances. A mathematical model of a ship, for example, would not have the graceful curves of marine architecture; it would, as likely as not, be the bowl of the Three Wise Men of Gotham – or even, as in one famous calculation by Newton, more like a sieve than anything else.

On the other hand, one ought not to use the word 'model' to mean just an abstract theory. It does, I think, denote some sort of mechanism or interaction, or articulation of bits of theory. It is fashionable nowadays, in sociology, or psychology, to talk in terms of models – 'Freud's model of the unconscious as the field of action of the id, the ego and the super ego', or something like that. In many cases this is objectionable, because one does not like the sense of the word 'model' diluted until it becomes synonymous with 'theory'. There ought at least to be a remnant, a reminder of the idea of a mechanically interacting system, and at least a suggestion that the system is amenable to mathematical analysis. Perhaps this depends on the intellectual sophistication of the speaker and his audience. I picked up a book on the philosophy of science the other day, where the author objected to the idea of curved four-dimensional space–time as a 'model' in the theory of gravitation because he found it difficult to picture. All I can say to that is that, having lived with this notion for 15 years or so, one can begin to feel at home with it, and tolerate it, and even take it for granted, just as some of us no longer feel completely bamboozled by Picasso's paintings.

That was meant as a warning against very vague, or perhaps against very formal, models only characterized by their equations. But even precisely defined models, easily pictured, have their dangers, for they may take our interest away from physics. We come to see the simplified abstracted model as if it were real. An example is the system known as classical hydrodynamics – that is, the motion of a fluid without viscosity – the elegant mathematics of which has been the basis of many a tricky question in the old Cambridge Mathematical Tripos. Unfortunately, almost no real system behaves like this. There is a well-known theorem of classical hydrodynamics that there should be no resistance to the motion of a sphere through such a fluid, which would be delightful for the aircraft manufacturers if it were the case – but it is not. There are other cases in contemporary theoretical physics of too much concentration on the mathematical properties of over-simplified models which lack significant features of the systems they are supposed to portray. Thus, many papers are published on 'one-dimensional crystals', or 'one-dimensional liquids' – that is, a row of atoms strung in a chain – with the pretence that the results obtained are somehow applicable to real solid or fluid three-dimensional systems.

Well now, suppose we have made our calculations, and they do not work, and give incorrect predictions; what then? We must cast about for reasons, and try to do better. We might discover, for example, that our mathematical techniques are too crude, although sound in principle. They tell me that it is only recently that it has been shown how the wind raises waves on the sea, although surely here all the separate equations have been known correctly for a very long time. Evidently this was a case of finding a better scheme of mathematical approximations, and not throwing away some important terms.

Or it may be that our model is too crude. An example – perhaps rather a technical one – is the so-called Einstein model for the heat motion of atoms in a crystal. Einstein proposed to treat each atom as if it oscillated in the average field of force of its neighbours. This leads to a very simple theory of some of the thermal properties of the solid. But the formula that one gets is not quite correct, and Debye showed that one could improve the answer greatly (especially at low temperatures) by taking account of the way in which the motion of each atom affects the motion of its neighbours – in fact, just as if there were sound waves travelling about in all directions through the crystal lattice. Debye was still using only a model, for he did not attempt to set up an exact formulation of the system, such as we might use nowadays with all the forces between the atoms; yet he got answers that are quite good enough for many practical purposes.

But there is always the lurking fear – or is it a hope? – that the formulation itself is unsound, and that we need to make some change in our assumed laws of Nature. Even in the best regulated of sciences this occasionally is forced upon us. For example, it was found, about 60 years ago, to be quite impossible to explain the structure of atoms by classical models; they were all hopelessly unstable; they tended to radiate away all their energy and then vanish, presumably in a puff of smoke, in a fraction of a second.

How does a new law arise? Here again we poach on the preserves of the historians and the philosophers of science. This is one of their favourite inquiries, to which they have bred up a number of answers. Far be it from me to try to bag any of these flighty creatures – especially the more ornate ones hatched from the eggs of psychoanalysis, and of Marxist sociology. Without prejudice against all such sport, let me suggest that what the inventor does is to look at a simple mathematical model with the known laws – and then he modifies the laws to see what happens. Just what leads him to think of doing such a thing is, of course, the subtle question in each case, but I think that is a fair and plain description of it from the outside. We might say that the inventor or discoverer or creator of a new law of physics 'toys' with the idea before he can demonstrate its validity. To make a shameless pun, let me call a model which contains new hypotheses, or new variations of old laws, a 'physical toy'.

There are some very simple examples of physical toys in the early history of the quantum theory. Max Planck was trying to explain the colour of the light that you see when you look inside a hot oven – what is mysteriously called 'black-body radiation'. He found that he got a very good answer if he assumed – quite arbitrarily – that the energy in the radiation could only be produced in finite lumps, and was not divisible into smaller and smaller quantities. This was the 'quantum hypothesis', which he embodied in his 'toy', or as we should now call it, the 'model' from which he calculated the spectrum.

Another famous example is Bohr's theory of the hydrogen atom, which is a model, based on the Solar System, containing an extension of Planck's ideas in the form of an hypothesis that angular momentum also had to be 'quantized'. Here is a 'physical toy' at its purest – very well defined mathematically, easy to analyse, and having just the properties that had long been known from spectroscopy. There is a robust realism about this model that makes it very appealing; and even though it has been superseded as a 'fundamental' interpretation, it is still useful as a 'picture' of the atom.

We might look at another well-known case, such as the wave theory of light which was developed at the beginning of the nineteenth century. The new hypothesis was 'light is wave-like, not particle-like'. There existed a good theory of waves in elastic solids, so it was not too difficult to write down the mathematics of such waves as if for light, and to get out some consequences, such as the slight bending of light round corners, and other observable diffraction phenomena. The point is that this was, in the first instance, only a 'toy'. Nobody knew what the waves were 'of', nor why they should be so. It took another 60 years before Clerk Maxwell showed that they were waves of electric and magnetic force, and brought them under more complete control.

Nevertheless, the toy had served its purpose, for with its aid a vast range of optical phenomena had been predicted, described, or explained. Indeed, it worked so satisfactorily that it became too realistic. We recall Lord Kelvin suggesting that if the waves were elastic waves in a solid 'ether', then this must be enormously more rigid than steel – yet solid bodies seemed to pass through it without hindrance. Because the only sort of waves which were easily visualizable to the nineteenth-century physicists were waves in elastic solids, they could not help worrying about all the other properties that seemed to be implied by the theory – they forgot that it was only an analogy, or, as we should call it, a model.

This sort of misinterpretation, ridiculous as it sometimes appears, is valuable if it does indeed bring out the contradictions in the model or toy that we are using. It may show clearly that it is only 'phenomenological' or 'heuristic' – very valuable words, these, that one can use either as knock-down weapons of offence, or as a shield against being accused of making inflated claims; I highly recommend them to any writer of dissertations, theses, scientific papers, referee's reports, book reviews, etc. They mean that it is specially designed to demonstrate certain phenomena, but is not universally valid. In fact, all toys, all models, all mathematical formulations in theoretical physics are ultimately 'only phenomenological', so one is quite safe to assert it of any particular theory.

We can see the status of the 'wave hypothesis of light' quite clearly, because we can look backwards to it, and then see where it led, and how it eventually was incorporated into a larger theory. But in our own day, in the theory of elementary particles, we see toy-making in its initial stages. There is a great deal of excitement this year about a theory that started off under the name of the 'Eight-Fold Way', and then became 'SU_3' and is now either 'SU_6' or 'SU_6 by SU_6' or U12, or some such incantation. As everybody knows, this is a scheme of classification of many of the particles which have been seen in high-energy physics – a scheme reminiscent of Mendeleev's Periodic Table of the Elements, but rather more subtle.

The heart of this scheme is that it depends on assuming that there are several different 'operators' (that is very big magic) that have mathematical properties closely akin to the properties of quantum mechanical spin. It follows (I mean that is what the theoreticians deduce from the axioms that they prescribe for these operators) that they only have two, or four, or eight, or ten, or whatever it is, different quantum numbers, one for each particle. When one has made the proper identification of particles in this scheme, one can predict their masses, the way in which they transform into one another, and so on, just as Mendeleev was able to predict the chemical properties of missing elements from his Table.

But, like Mendeleev, we do not know the meaning of the scheme. We do not know what the operators signify – we call them quaint names like 'strangeness' or 'baryon number' or 'isotopic spin'. We do not really understand what they operate on, just as we did not know what was 'undulating' in the 'luminiferous ether'. Perhaps one day we shall solve this puzzle, and our toy will have become no more than a model.

A physical toy may, indeed, be just an equation, such as the famous wave equation of Schrödinger, which even chemists find they have to know. This is an equation for something – we call it the 'wave function' or simply label it by the Greek letter ψ – which is related in a very curious way to the presence or absence of an electron in the neighbourhood. But what Schrödinger himself showed was that the solutions of this equation agreed with the energy-levels of a hydrogen atom and that was enough to build a whole theory. It was only later suggested that the square of the wave function was really a probability density. Schrödinger could, so to speak, toy with the equation, and believe that it had some physical significance, before he had any clear idea of what the variables meant, or how the equation was to be derived from more abstract laws.

From these examples it would be rash to draw any conclusions about the 'pattern of discovery' in theoretical physics. But we can say something about the role of pure mathematics itself in this mysterious process. It is obvious that we want to make our models, and the laws that we assume for them, as general and abstract as possible, so that the process of looking for alternative laws, or of modifying the model in some significant respect, can be kept under control. We want a systematic procedure, or at least a rational scheme of procedures for our research. Pure mathematics does just this. It tells us exactly what the minimum rules are if we are to achieve a certain result. It is a study, not only of models that we have already formulated out of physics, but of hypothetical models in which some of our assumed rules have been relaxed.

The best-known example of this is the geometry of space. As everyone knows, Euclidean geometry depends on Euclid's axioms. Relax one of these – the parallel axiom – and we get either of two special sorts of geometry: hyperbolic geometry or elliptical geometry. Relax another axiom or so, and we get Riemannian geometry. Relax some more rules, and we get affine geometry. This is as far as my knowledge carries me – but no doubt there are still simpler, more general geometries of which the list of axioms is even shorter.

Now Einstein, in his search for a more general theory of relativity, needed to get away from Euclidean geometry, which did not seem consistent with physics; he had already available, in this branch of pure mathematics, many theorems of Riemannian geometry, which, in fact, gave him the model he needed for his theory of gravitation.

From this famous episode, also, we can draw another lesson. It will be remembered that the theory of gravitation was preceded by an earlier theory called the special theory of relativity, which explained some of the facts that had been discovered about the way light was propagated, and what happened when physical objects travelled nearly as fast as light. Shortly after this theory was formulated (I do not propose to argue whether Einstein did more or less than others in this) Minkowski showed that it could be derived mathematically, in a particularly simple way. He pointed out that it ought to be considered as a problem in four dimensions, rather than in three-dimensional space plus one-dimensional time. Minkowski did not add anything to the physics; all the results that he derived by his method were already known from Einstein's arguments. He was only looking at the same model through slightly different mathematical spectacles. Nevertheless, this way of doing the mathematics turned out to be very powerful, and could easily be generalized, as it was by several people, until eventually Einstein took the next step of incorporating it into his gravitational theory.

Here we see the role of the 'formalist' – the type of theoretical physicist who spends his days trying to simplify, axiomatize, or otherwise regularize, the models invented by other people. In spirit he is something of a pure mathematician, for rigour is essential to his outlook, and he may really care nothing about the physical system of which his axioms and theorems are supposed to be a model. We must be careful not to despise the formalist, for his work is not always sterile: the clarifications that he achieves, the simplifications of technique, of notation, of formulation, may be necessary prerequisites for the next leap forward.

There is a striking example of this on the very highest plane. The great modern theories of quantum mechanics, the work of Schrödinger and Heisenberg, sprang up at Göttingen where Courant and Hilbert had just published a book called *The Methods of Mathematical Physics.* This book contains most of the mathematical theory which is needed in quantum mechanics; it is still one of the best texts on the subject. Yet it was meant to be a formal study of mathematical physics as it then existed – that is, of the mathematics of classical physics, such as elastic waves, optical diffraction, and so on. It was, if you like, what a couple of pure mathematicians made of the brilliant and immensely inventive work of Lord Rayleigh, who, 40 or 50 years previously, had used elegant, effective, but rather specialized arguments, based on simple models, to explain a wealth of phenomena about the propagation of sound. But the book was written with such generality, with such insight, with such simplicity, that it provided the mathematical schemata for a wholly new departure in theoretical physics. It is a very nice point whether it was just good fortune that the right sort of mathematics happened to be available when quantum theory was striving to come to birth or whether it was inevitable that such a beautiful and all-encompassing mathematical theory must transform our vision of the state of things into its own likeness.

The moral of this is that a training in proper pure mathematics of some sort is essential to the good theoretical physicist. It is not merely a matter of knowing all the formulae for solving a particular equation; it is the ability to see, or experience in seeing, a given model, or a set of laws, as one of a whole class of models, of more general and abstract properties. It is the ability to see into the fourth dimension, to recognize the properties that are invariant, to watch for the still centre in a kaleidoscope of transformations.

This ability, or this training, must be balanced with 'physical intuition', which is experience of the properties of all sorts of models, a feeling for the way they will behave, and readiness to construct or adapt a model or a toy to suit the problem on hand. This quality is much less definable than the first, which is essentially that of the pure mathematician; but it is the heart of the matter. I suppose it has to do with imagination, or 'divergence' of personality or something. There is, indeed, a tension between the 'formalist' and the 'intuitionist' which may be essential. It may not be possible for a single personality to contain the qualities needed for the invention of models, on one hand, and for the rigorous investigation of their properties on the other.

Nevertheless, the disciplines of mathematics and of physics must somehow be combined to produce what we need. It is no accident that the titles 'theoretical physicist', 'applied mathematician' and 'mathematical physicist' are so confounded and that the persons who bear them do not know in which camp to pitch their tents. But that is a problem of academic tidiness, which is surely the antithesis of philosophy.

To show that there are other views on the proper ways of looking at the world, let me quote from the autobiography of a famous eighteenth-century Italian historian, Gianbattista Vico, who says of himself:

> When he had discovered that the whole secret of the geometric method comes to this: first to define the terms one has to reason with; then to set up certain common maxims agreed to by one's companion in argument; finally, at need, to ask discreetly for such

> concessions as the nature of things permits, in order to supply a basis for arguments, which without some such assumption would not reach their conclusions; and with these principles to proceed step by step in one's demonstrations from simpler to more complex truths, and never to attain the complex truths without first examining singly their component parts – he thought the only advantage of having learned how geometricians proceed in their reasoning was that if he ever had occasion to reason in that manner he would know how.

So much for mathematicians and logic; and in another place he says:

> A short time after this he learned of the growing prestige of experimental physics, for which the name of Robert Boyle was on everyone's lips; but profitable as he thought it for medicine and spagyric, he decided to have nothing to do with this science. For it contributed nothing to the philosophy of man and had to be expounded in barbarous formulas . . .

10

Not so much a Model as a Theory*

There's something gentle and wholesome in the word *model.* It conjures up an image of toytown, of miniature railway engines, perfect in every detail, complete with brass alarm bells and firemen's shovels, puffing and pounding around the track. Yes, it must be a *working* model, like one of those marvellous engines in the Science Museum that won my heart as a child. Thus and thus is how it's really driven – at the touch of a button the crankshaft rotates, the camrods go up and down, the valves open and close, the steam goes in here and pushes until the piston gets to here, and the cycle is complete.

The word *model,* that sweet, innocent, helpful word, is carried over into our science. The Victorian physicists wanted models of the propagation of electromagnetic waves in the ether, pushing and pulling, cranking and reciprocating like one of their lovely steam engines. And then there was J. J. Thomson, with his model of the atom, like a currant bun – negative electrons embedded like raisins in a dough of positive charge. And Rutherford, who saw that it was a planetary system, with the electrons in orbit around a nuclear sun.

That's how we like our models in science – not just dry mathematical equations but living metaphors of the world we know. We give them the most homely names: a nucleus has its 'liquid drop' model which holds itself together, and bellies and wobbles, just like the real thing. Or, on another occasion, the very same object is likened to a 'cloudy crystal ball', more mysterious and vague, diffusing and scattering radiation that falls upon it with characteristic lack of definition.

The very beauty of a theoretical model, with its own internal precision and consistency, attracts us irresistibly. They say that every scientist, like Pygmalion, falls in love with his own model. There are theoretical physicists who spend a lifetime perfecting and polishing some simplified

* BBC Radio broadcast, 28 November, 1976.

version of an atom or a star, calculating to ten places of decimals, with the same devotion as the retired bank manager running his railway network round the garden to the inexorable discipline of Bradshaw's timetable. Whole treatises are written on the theory of one-dimensional models of crystals, or of lattice models of liquids, which never existed, never could truly exist, except in the human mind.

I speak of the power, of the seduction, of a word. For if you substitute the word *theory* for *model,* how much less certain and constructive seems your enterprise. After all, scientific *models* are only *theories,* coherent, self-consistent, readily comprehended theories, but no more attached to reality than verification or successful prediction can ultimately justify. But a theory that has been shown to be invalid vanishes from contemporary thought – whereas a compelling model, however false to the truth, can live on forever.

That's why, in the full glory of his conceit, many a psychologist or sociologist speculating on some vaguely-articulated and ill-defined propositions concerning human behaviour, dignifies himself as the engineer of what he describes as a 'model of', say, 'cognitive dissonance' or 'the sociodynamics of identificatory fields'. He wants to create a permanent ideal in the Platonic world, a system whose little wheels will go round, whose camshafts will jiggle, whose valves will open and close, for ever more. He will be deeply grieved and insulted if you tell him the truth – that he has put together a set of hypotheses, that his theory is a stimulating speculation, that his conjectures, though by no means uninteresting, wait to be verified. Such words, full of scepticism and doubt, hinting at darknesses and loose ends, quagmires of imprecision and feet of illogical clay, are to be banned from academic discourse, like the smell of death amongst the living.

But without our armies of models – logical infantry, empirical cavalry, mathematical tanks and guns – we couldn't play our intellectual games, our battles across the nursery floor of Academia. What are these models of ours but charming toys that we may bring out of the cupboard for game after game of competitive, creative play. It's only when we come to speak, and think, of life and love and things of the heart and spirit, that we must not be deceived by mechanical metaphors – by models that are too sharp and simple to represent the clouds, the perfumes, the music, the voices, the symbols, the visions that are the true stuff of reality.

11

Words and Symbols*

The peculiar problem with physics is that it cannot be stated entirely in words. Physics is a science in which we attempt to represent the phenomena of nature in mathematical form. Without the machinery of number and quantity, symbol and relationship, formal operation and logical necessity, it would never have penetrated the atom or the galaxy. Writing up physics in dictionary language is like trying to describe a picture to a blind man.

For those of us who attempt to explain science to the general public the problem is all too obvious. We are forced into desperate devices of metaphor, we skim lightly over vast areas of numerical data and algebraic analysis to give an impression of what we would be after. Mathematical argument is too strenuous and specialized for popularized science and we must simply do without it.

The interesting question, however, is whether mathematics should entirely dominate the exposition of physics in the technical scientific literature. Many theoretical physicists seem to think so. Their scientific papers, review articles, technical reports and textbooks consist of strings of formulae linked with mechanical phrases like 'it follows that' or 'substituting in equation 7.77 from equation 8.88 we get . . .' or (when the logic of the next step is a bit dicy) an appeal to the good sense of the reader with words like, 'obviously' or 'it is clear that . . .'

Up to a point, such style of exposition is honest and unambiguous. When he cannot follow the text with full comprehension, the reader may himself take up paper and pencil and, if he has the mathematical dexterity to jump a few gaps of algebraic manipulation, then he will soon have the inestimable pleasure of participating in the thought processes of the writer and being carried along to the same conclusion. Such writing is like a blueprint, incomprehensible to the layman, which the skilful architect

* BBC Radio broadcast, 19 December, 1976.

or engineer of mathematical physics can read constructively out of his own experience. To the expert there is real pleasure in grasping the inner harmonies of a well-expressed mathematical argument.

But of this style of exposition I often get the impression that the author lives entirely within the abstract world of his own symbols. Mathematically speaking, it may be quite sufficient to list their meanings as one might for a computer program. But for human use, they need a more lengthy and courteous introduction than a defining phrase such as 'where *R* sub zero is the radius of the protogalaxy at the moment of star formation' reinforced from time to time by repetition of the verbal label 'the protogalaxy radius, *R* sub zero' in case the attention fades, or to avoid confusion with some other symbol of very different meaning.

However compelling, however rigorous, a formal mathematical argument is not self-sufficient. Physics is about natural phenomena. The challenge to the theorist is to establish the metaphorical parallel between the real world of perceived material objects and the ideal realm of mathematical models. To keep in touch with electrons and galaxies, magnetic fields and ocean waves, quarks and quanta and quasars, we need *words,* we must give them their own true names. Abstract symbols, x and y, alpha and omega, plus, minus and equals, have no permanent homes of their own in the human mind. To *understand* physics, to feel its concrete necessities, we must accept the material implications of grammatical sentences containing nouns and verbs, the words for things and actions. Einstein's equation, $E = mc^2$, carries little weight until we have said it aloud to ourselves, 'All mass is equivalent to energy; all energy has mass', and listen to the reverberations of these statements within our own understanding.

In writing up or talking about physics, we must keep a thread of verbal reformulation and interpretation running in parallel with the mathematical formulae and establish linkages crossing and recrossing the boundaries between these two modes of communication. Yet the two modes should not be intermingled too closely, with equations and sentences jumbled together into a continuous text. Even for the mathematical physicist, verbal statements and symbolic relations do not occur within the same regions of the brain. We must beware of falling into a pidgin language where word and symbol, object and concept, are crudely compounded.

To convey the full meaning of physical thought, we need yet another mode of communication – visual imagery. 'One picture,' they say, 'is worth a thousand words' – and can convey to the intuition the essence of a whole page of algebra. The innate capacity to manipulate and transform a diagram or map is one of the glorious powers of the human mind. But a map is useless unless its principal features are named. And there can be a great step forward in our understanding when we put a new name to a newly observed feature on such a map. Whatever other media of communication, whatever other languages we may bring to our aid, we cannot dispense altogether with words, with plain English, with didactic discourse, according to the rules of grammar, in telling each other about the physical world.

12

A Question of Upbringing*

Students should be taught to appreciate the unity of theory and experiment in physics.

At the recent Discussion on the 'Principles of Science' Tripos, many well-bred dons may have been embarrassed to overhear a family quarrel amongst the physicists. 'Physical thinking is essentially mathematical', said Dr Pippard. 'A large part of physics can be explained in words', said Mr Ratcliffe. The controversy is not simply about the problem of teaching physics to humanists, nor is it just one of those deep philosophical questions which all depend on what you mean. It is concerned with how to teach physics to bona fide students of the science.

My own prejudice, as a professional theoretical physicist – ah, there's the nub – is the existence of two species in the genus 'physicist': 'theoreticians' and 'experimenters'. Is there something special about physics, which makes sense of this separation of function, comparable, shall we say, with the distinction between surgeons and physicians? Is there a real division of the labour of research between two types of person, mathematical and practical, or is this an artificial, too-rigid classification indicating too-narrow specialization?

The implication of the dichotomy is that research proceeds in two separate, but complementary, steps. E, the man in the white coat, makes an apparatus of black boxes full of electronics, vacuum pumps, magnetic coils, etc. and, exercising great manipulative skill, measures some significant physical quantity. The pointer readings he plots on graph paper; through the points a curve is drawn, and published in a paper on *The Thermophilic Coefficient of Technetium.*

This paper is seen by T, the man in the white collar. He skips the sections labelled 'Experimental Procedure', but pauses when he reaches the curve, mutters 'Now I wonder if . . . ?', reaches for his fountain pen and proceeds to cover sheets of paper with x, dz, α, dy/dx, etc. He appeals to Relativity, to Quantum Theory, to the Calculus, and all the parapher-

* Only a typescript survives; probably published in the *Cambridge Review* about 1956.

nalia of higher mathematics. His paper (*The Theory of the Thermophilic Effect*) ends with another graph labelled 'Theoretical values of . . .', and the remark that: 'The result of the present calculation is in good agreement with the experimental results (E, 1957).' If T is clever, he may add: 'It would be interesting to study the frigorophobicity of modernium, which should, as may be seen from equation 6.1.3(7), have a λ-point near the characteristic temperature.'

E also keeps up with the literature. In a review paper by R he sees a reference to the paper by T. He fetches the journal from the library (or finds an inch-thick stencilled 'preprint' in the dusty pile in the corner) and tries to read it. The pages of symbols are flicked over until he comes to this 'prediction of the theory'. The newest research student is called in, and set a-measuring. And thus one cycle of scientific progress is completed – one egg laid, one chicken hatched and grown to pullethood.

This is a parody of the system, but this is how it often works. Neither side understands, nor cares for, the work of the other. Experimental papers at conferences are boring; theoretical papers are incomprehensible. The result of such failure of communication is usually mediocre science. The separation of functions is entirely artificial. Theory and experiment, measurement and mathematics, hands and brain, are too intimately one to be separate professional activities. Without venturing onto philosophical ice, we can agree that physics uses mathematical symbols and abstract concepts to make meaningful statements about the results of observation. The experimental evidence for the theory is a part of the theory; the theoretical analysis of the experiment is a part of the experiment.

Of course there is natural cause for the division. There have always been physicists with a flair for experiment – Faraday, Joule, Rutherford – and those with a genius for mathematical analysis and interpretation – Newton, Clerk Maxwell, Einstein. It is no more surprising that these have concentrated on one or other aspect of the science than that some historians prefer searching attics for ancient charters, whilst others meditate on the general causes of historical events. Most of us are specialists in our work, because we have not the time, or energy, or ability, to do all things well. Yet there have been some who seemed entirely ambidextrous. Enrico Fermi will be remembered as much for his entirely theoretical invention of the neutrino as for the construction of the first atomic pile. Even the two antagonists in the recent Discussion, Mr Ratcliffe and Dr Pippard, could scarcely be labelled precisely as E, or T, from their published papers.

The misfortune is that, because scientists have shown tendencies to specialize in these two ways, the notion has grown that students, when they enter the University, should decide to which species they belong, and be trained exclusively to that end. In the Natural Sciences Tripos a man must do many hours of practical work, and must read three experimental sciences, with only a grain of mathematics; in the Mathematics Tripos he will start from a set of God-given equations (said to have been 'discovered' by Newton, Einstein or Schrödinger) and learn to manipulate and solve them, without care for their origins, or consequences, or relations with the world.

Most people now agree that such a rigid partition is wrong, and that physics should be taught more as a whole. Nevertheless, there is a difficulty in practice. The theoretical aspects of physics are mathematical, and mathematics is hard to grasp. Each one of us has a mental barrier, a realm of abstraction into which we cannot enter, however hard we try. The effort of concentration becomes too great, and the concepts become meaningless. The barrier is in a different place for different people. Must our average practical man, good with his hands and ingenious with devices, be forced up to this limit of comprehension, instead of being taught more of his craft and technique? Must the brilliant mathematician, anxious to sharpen the edges of his mind, and cut his way into new knots, be blunted with brute facts which he could easily learn later if he should need them?

In the Discussion, Mr Ratcliffe gave the impression that there was, at least for the practical man, a way round. The thought processes of physicists could be verbalized, whilst the mathematical part of the argument could be isolated inside something like a very clever electronic computer, which, when fed with 'input' equations, automatically produced 'output' equations. These parts of the theory were to be taken on trust by the experimentalist and need not be taught to him.

This is, up to a point, true, yet it is misleading. There are certain steps in the logic of a physical argument which depend only on formal mathematics – the evaluation of an integral, the solution of a differential equation, the diagonalization of a matrix. These are mechanical manipulations which, even in a 'theoretical' paper, are often put into an appendix (though even there, an experienced theoretical physicist would use his physical intuition, as Mr Ratcliffe himself pointed out, to make sure that he had not made a mistake in the calculation).

But most of the verbalized concepts of the physicist are also essentially mathematical. A symbol does not create an abstract notion; it only makes it easier to deal with. The ideas of *momentum*, of *energy*, of *entropy*, of *electromagnetic waves*, or *fields of force*, of *spectra*, of *quanta*, of *spin*, of *parity*, etc., etc., were all born as mathematical concepts, but have grown into physical entities, which we are taught, as physicists, to grasp intuitively. The puppets are manipulated so vividly that we clothe them with flesh and blood. Although, when we use them as physicists, we think these entities are as natural as billiard balls, we only reach this stage of understanding by working through their mathematical derivation and feeling the invariant physical objects coming to life under the symbols. To accept abstract entities on trust is the way to a world of phlogiston, caloric and the luminiferous ether.

This argument also has its converse. The mathematician may be tempted to solve the equations and miss the essential point. Because he has not learnt enough physics to 'see through' his algebra, his calculation may become planless, random and useless as a guide to experiment. The feel for the 'physics' of a problem is a vital part of his mathematical equipment.

The solution of the teaching problem is not simple, but I think one can discern the general principles. If we pretend to teach physics to Natural Scientists or to Mathematicians, we must teach it as one science, from a unified standpoint. A theory must be presented both as a set of equations and as a pattern of phenomena. We must explain what we believe, as precisely (therefore, often, as mathematically) as we can, and also why we are forced to believe it. This must mean more mathematics for some students – and more glimpses of the real world for others. But this discipline should be tempered with another principle – that we are not all born equal in our power to assimilate the particular in the general. We must be ready to give a range of courses, fast and slow, more and less abstract, more fundamental for some, more concrete and practical for others.

13

Do some Theories Stink?*

The real danger is not ideology in science, but scientism in ideology.

Does it matter whether science is 'really' ideological? This could be one of those highbrow philosophical questions, to which we might apply Sam Johnson's blunt comment on idealism: 'I refute it thus', he said, kicking hard at a stone until his foot rebounded. The practical achievements of scientific technique are too concrete to be disturbed by political or ethical wishful thinking.

But at a recent conference on The Social Impact of Modern Biology, a paper by Dr Robert Young, 'Evolutionary Biology and Ideology', was greeted enthusiastically as an important contribution to the general theme of social responsibility in science. His thesis was that 'it is routinely impossible to distinguish hard science from its economic and political context and from the generalizations – which often also serve as motives for the research – which are fed back into the social and political debate.' With characteristic seriousness and learning, he propounds this argument, not as an academic exercise but as a guide to behaviour. He tells us, in effect, that only by recognizing the inevitable ideological foundations of scientific thought can we correct its bias and apply it properly to social problems.

In itself, this paper exemplifies the interaction between the scholarly, the ideological and the practical that he has set out to discuss. If we believe, as he does, that the sociology of knowledge demonstrates that 'there is no essential distinction between the scientific and the evaluative', then we shall attempt to do research and apply scientific discoveries in other ways than we should if we believed that science is basically objective and independent of human values – just as we might be strongly influenced in our political decisions if we believed that it had been scientifically demonstrated that black people were mentally inferior to 'pinko-greys'. To that extent, one must treat Dr Young's argument as seriously

* BBC Radio broadcast, published in *New Humanist* **88**, 150–2 (1972).

as he does; scientific, sociological and philosophical opinions are by no means irrelevant or impotent in the affairs of men.

But I certainly do not agree that the relationship between scientific knowledge and ideology is so close as to merge into identity. If we did not normally make a very sharp distinction in our minds between scientific and ideological arguments, then a great deal would be lost. Selling science short is this year's intellectual fad, but it is not quite fair to spread it around that the old firm is totally morally bankrupt. I may sound a bit like a company spokesman complacently denying rumours of insolvency, but I think we still have a few capital assets, an experienced work force and a lot of cash in the bank, despite the loss of good will. Actually, the situation is rather interesting, as a general problem in intellectual accountancy, in ways that are not always obvious from the official balance sheet of material profit and spiritual loss.

The first thing to say is that Science, with a capital S, is much too complex to be treated as a single 'thing' to which we may apply simple epithets like 'good,' 'false', 'productive' or 'bourgeois'. It is a large-scale human activity of many aspects and many intersections with other components of society in the intellectual, material, psychological and political dimensions. The sociology of knowledge – the analysis of the relationship between the theoretical content of scientific treatises and the writings of political philosophers, novelists and other guardians of written culture – is only one such intersection. We must also take account of the psychology of scientific discovery, of the internal workings of the scientific community, of the effects of technical change, of economic forces, of industrial organization, of education, of war and of peace. To damn science for being tied to the wrong political ideology does not seem to make much more sense than the corresponding generalizations which condemn it for being limited by the finite powers of the human mind, for being just a game played by scientists for their own amusement, for being a soulless product of dark satanic mills, or for failing to fit into a tidy filing cabinet at the Ministry of Technology. It is too deeply embedded in our thinking and culture to be dragged out like that by the roots. As Galileo muttered, under similar circumstances: 'But it really does move!'

So there are many different ways of trying to be socially responsible in science – and to science and for science and by, with or from science. At every point of contact between the realm of knowledge and the realm of action, we must learn to use imagination, sympathy and good sense. Abstract ideology does not provide an adequate decision-making formula for assessing the likely consequences of a new chemical process, or the human benefit to be gained from a particular line of medical research, or even such bitter dilemmas as those that arise in modern war. Belief that one is on the side of the angels can strengthen the will but does not automatically justify a cause nor make it conform to hard reality.

To discuss this question at all, it is not enough to point to some generalized philosophical necessity. Of course, we are the product of our times and have learnt to use our minds in the social context of our own generation. 'Our conception of reality itself' is, as Dr Young says, 'socially constructed'. So what? Can we then lift ourselves up by our own mental bootstraps or add a cubit to our intellectual stature, and look down upon the vulgar mob with superior insight? There is no way out of this vicious circle, except to self-mockery or mysticism.

The task is to discover the characteristic features of scientific knowledge within this human straitjacket. We used to believe in an elixir called 'scientific method' which, if sipped in the right frame of mind, would permit us some glimpses into that ideal world named 'truth'. Alas, this comforting positivist doctrine is no longer tenable; but that does not mean that science has lost its virtue to Demon Unreason. As the dream of grounding our understanding on strict fact and firm logic has faded, we have begun to see the importance of the scientific community itself as a mechanism by which the validity of science is attested and maintained. Science, in other words, is characterized by what I call 'consensibility': it is a body of observations, principles, theories, etc. which has no higher standing than that it has been exhaustively examined and freely criticized, by a large body of people, until an overwhelming consensus of agreement has been reached. The subject matter of science is thus severely restricted: it includes the theorem of Pythagoras and the fact that water expands when it freezes, since no man on Earth has ever been able to persuade others to the contrary; it does not include the ideal recipe for a sponge cake, nor any opinion on the novels of D. H. Lawrence, since these are matters on which the best trained tastes may properly differ.

At first sight, this characterization seems to be deliberately phrased to isolate science from 'ideology' in the general sense of moral values, political opinions, etc. This is, indeed, the old-fashioned positivist doctrine on the subject, against which Dr Young and his fellows are reacting with such passion. Ideology, we might have said, is either as pervasive and invisible as the air we breathe, or else it is a matter of such controversy and disagreement that it could never be the foundation stone of a consensus. Immediately we become aware of such a flaw we must do our stern scientific duty, and cast out the defective link in our argument. As Lagrange replied to Napoleon, when asked if he believed in God: 'Sire, I have no need of that hypothesis.'

Alas, such intellectual arrogance is no longer acceptable. Having lost the anchors of 'scientific method' and the 'principle of induction' we are adrift on that raft of interlocking, trussed-up bundles of contradictions and misconceptions – the group of human minds constituting the contemporary scientific community. Scientific truth turns out to be no better than the corporate wisdom that this group can arrive at, in the prevailing conditions. A wind of delusion, a strong current of ideology – much the best example is medieval scholasticism – can carry the whole scientific community, and all our scientific knowledge with it, into very distant seas, towards very strange lands.

This, in fact, is Dr Young's essential argument; but I would go much further. Ideological prejudices are the source of only a small proportion of the errors and fallacies embedded in any branch of scientific knowledge. In some ways, they are the easiest to correct, for they are continually subject to public dispute. In the hurly-burly of life, we encounter a mish-mash of contradictory opinions on morals, politics, aesthetic values and religion, which warn us of the danger of a one-sided view and largely cancel each other out. It is much more difficult to detect the inconsistencies within a generally accepted body of scientific thought, reinforced by the combined and supposedly dispassionate opinion of the experts themselves. 'Phlogiston' and 'caloric', 'spontaneous generation' and 'the inheritance of acquired characteristics' were all, in their day, taught as truths of official science. The 'values' with which they were laden had little to do with ordinary social life; they stemmed from the authority of those great scientists who had discovered, or failed to controvert, them in a previous age. Old wives' tales have nothing on the strange doctrines to be found – with the wisdom of hindsight – in the most solemn and majestic of ancient scientific treatises.

But the pursuit of these and other fallacies is the professional *art* of the theorizing scientist. He knows of phenomena, of observations, of experiments whose interpretations are confused, disputed, mutually contradictory or incomplete. His task – the very essence of his scientific 'creativity' – is to build a new pattern of concepts and explanations that is more complete, more consistent and thus more acceptable to his colleagues. It makes no difference in principle whether the defects that he is trying to mend are 'ideological' in the general social sense or whether they come from within the scientific consensus itself; his job is to criticize and to correct errors of all kinds.

To do this he needs what are nowadays called 'models' – generalized concepts fitted together into mechanisms that bear some resemblance, in structure and behaviour, to the material system on which he is doing his research. If his choice of models is limited by his ideological preconceptions, that's too bad; it may rob him of a famous discovery. But he is not bound, by any of the rules of his game, to look for his models in any particular philosophical system. His theory will be judged to be acceptable by its power to explain or predict the results of experiment within the scientific context, not on whether it happens to be drawn initially from the stock in trade of a 'good' ideology.

A successful theoretical model may be assembled from very diverse parts. J. J. Thomson tried to show that the electrons in an atom were embedded in the positive charge like currants in a bun; Rutherford and Bohr guessed that they were like planets going round the Sun. Newton thought light was like a hail of bullets; Young showed that it was more like the waves on a pond; nowadays we think of bullets again, but spinning as they go. A nucleus behaves like a drop of liquid – and a neutron star is just an enormous nucleus. Is crystallography a military science because the orderly rows of atoms remind us of regiments on parade, or piles of cannon balls? We cannot think about *anything*, except by analogy and metaphor. Scientific thought, no less than primitive thought, is made by the process that Lévi-Strauss called *bricolage;* out of *bric-à-brac* – miscellaneous objects selected from the rubbish bins of our memory – we make beautiful, useful, new toys and machines.

For this reason, Darwin cannot be accused of bourgeois bias (or something!) because he made brilliant use of the pessimistic social doctrines of old parson Malthus. What Darwin needed was an idea, a model, for selection by competition; by chance he got it from that particular source. But once he had seen the point, and had applied it to animal and plant populations, he had made it part of the general theory of biology. It could perfectly well have been noticed or invented by any sufficiently perceptive genius from Aristotle onward – except that the question had not yet been asked to which it was the answer. From then on, for Darwin and his scientific colleagues, natural selection could be studied as a biological mechanism, quite apart from any political or sociological misapplications that might be made of the same basic idea.

Patchwork theory

The fundamental fallacy is to believe that a theorem, a hypothesis or a readily verifiable observation can have an unpleasant odour because it happened to be born in the mind of some cruel, foolish, selfish or otherwise undesirable person. The first principle of natural philosophy is brief enough to be phrased in Latin: ***hypotheses non olent***. 'Theories don't smell.' Scientific concepts are not guilty by association, nor do they come to us 'trailing clouds of glory'. Entering the scientific domain, they shed their load of 'values', and are free to be just themselves.

The tragic history of Lysenkoism proves this point to the hilt. Now that some of the inside story has been revealed, it is more than ever clear that political tyranny was used to impose the nonsense of a half-educated crank upon the Russian scientific community, wasting vast resources and doing great personal cruelty to many men of wisdom and integrity. The fact that this was done in the name of a 'good' ideology does not, somehow, justify or mollify such actions. On the contrary, those of our own scientists who lent themselves to Lysenko's ideological criticism of 'Mendelism' showed precisely their own lack of grasp of the basic principles of science itself. A body of knowledge which is supposed to contain some gross error for an ideological reason may certainly need to be criticized and cured – but it cannot be corrected simply by substituting the supposed consequence of the 'correct' ideology into the argument. The possibility of an ideological bias may well be suspected – but it has to be demonstrated by the usual methods of science, within the framework of consensibility, by painstaking research and effort.

Yes – *effort*. Some clever people without experience of research seem to have no idea how difficult it is to get anywhere near the truth, even in a mathematically rigorous, unspeculative science like solid state physics. Research is an immensely laborious activity and knowledge only progresses by the combined exertions of many many people, over many many years. Looking back, we tend to see only the path of success – the great steps forward, the brilliant leaps of intuition; we ignore or forget the errors that have been corrected, the data that have been discarded, the blind alleys retraced, the proofs that failed and the speculations that were fruitless.

Nor is all scientific knowledge as reliable as we pretend in our textbooks and encyclopaedias. Much of what passes for established theory is a patchwork of vague, scarcely consistent notions, cobbled together and stretched to fit a few scattered observations. All too often, we claim a victory when we have arrived at a satisfactory practical technique by trial and error. Because experiment and observation have no meaning without some sort of theory, we need at least provisional models, verifiable or falsifiable within a restricted range of phenomena. But our failure to falsify such a model does not always make it true. The enthusiastic research worker can be captured by his own metaphors, which expand in his imagination until they fill the whole sky. A whole generation of scholars, can be trapped in its own paradigms: pretty little theories, that explain a small class of specially contrived phenomena, are elevated into divine revelation concerning the nature of the Universe.

The message of science to the eager social reformer is much less firm than he so ardently desires. In fields such as psychology or sociology, where nothing is sharply defined or measurable, where concepts are plastic and everything is closely linked, the only sensible attitude of mind is extreme scepticism about all theoretical constructions and their logical consequences. It seems to me, for example, that an attempt to explain the varieties of human behaviour with the sole aid of Freudian psychoanalysis is about as hopeful a venture as setting out across the Atlantic in a coracle. Again, there is no objection to saying that some types of behaviour are 'intelligent'; but this does not justify the notion that a man has a single-valued, measurable, permanent attribute called his 'intelligence' – a platonic companion, no doubt, to his immortal soul! In the sciences of man, there are vast tomes of fascinating observations, and many local regions of intellectual order, but there is not the mathematically rigorous structure, based upon universal laws, that we find so compelling in physics.

Intellectual responsibility

The real danger, then, is not ideology in science, but scientism in ideology. Herbert Spencer's attempt to base a political economy on Darwinism; Darlington's views on racial inequality; Ardrey on aggression and social conflict – these are scandalous because they claim scientific credentials for mere opinion or prejudice. Speaking as if with scholarly authority, they take up models appropriate for certain ranges of biological phenomena and illegitimately transfer them to the human social arena. I agree fully with Dr Young that it is the social responsibility of scientists to denounce such crude impostures.

Yet I do not think it is our job to forbid altogether the attempt to discuss the human condition in scientific terms. Within its own sphere, natural science is the most reliable knowledge that we have, but that sphere is very limited. There are many other aspects of life and the world that demand some sort of analysis, to clarify our thinking and our action. Public debate of issues associated with racial differences, or with human aggression, is not necessarily irrational or ideologically committed, even if it cannot be kept within the boundaries of a mature scientific consensus.

The fact is that speculative diversity, disagreement on basic axioms, personal controversy, negative criticism and general intellectual disarray are as characteristic of science itself, at the active research frontier, as they are of ideological debates. At the formative stage of a science, it may be impossible not to use the style and language of ideology, while the attempt is being made to separate consensible elements from the chaos of facts and opinions. And, just as science has borrowed valuable metaphors from religion and politics, so there can be legitimate use of scientific models to illustrate social thinking. The criteria for success in this activity are not that we arrive at the answers preordained by some well-thumbed ideology or that we satisfy ourselves that we have found the answer to the Riddle of the Sphinx, but that by sensitive deployment of such analogies, tempered with much scepticism and doubt, we may see some of our older problems in a new light, from another aspect, with deeper insight. That is the meaning of *intellectual* responsibility in society.

14

What's in a Name?*

Scientists fall in love with their own creations – the words they have to make up for what they find in Nature. In 1834, Michael Faraday needed words to describe what happens when an electric current flows through a chemical solution. Not having had a classical education, he consulted Dr William Whewell of Trinity College, Cambridge, the acknowledged expert on naming things in pseudo-Greek. The geologists had come away from Whewell well supplied with words like *Pliocene, Miocene* and *Eocene;* he would be happy to oblige the distinguished physicist. In a delightful exchange of letters Faraday describes the phenomena and explains concepts which he wished to express in the archaic abstractions of artificial Greek. Whewell offered several alternatives, including some barbarous tongue-twisters, such as *skaiostechion* and *cetazetode,* that would have killed the science of electrochemistry on the spot. Fortunately, Faraday was a man of natural good taste and agreed with Whewell's first preference for those beautiful words *anode* and *cathode,* which are perfectly good Greek for 'the way up' and 'the way down', through which the mysterious particles, the *ions,* would pass. Now, after a century of physics and electronics, plus several extensions, generalizations and metaphors, we have in almost every home a cathode ray tube, on which magical shimmering pictures come and go.

Modern science has no time for such high-mindedness and linguistic delicacy; the names we get now for new scientific concepts commend themselves chiefly by their whimsicality or vulgarity. Yet the christening is occasionally graced with vitality and wit. Back in the 1950s, the big machines of high energy physics were producing whole families of elementary particles – dozens of them, interacting and decaying into one another with astonishing facility. Amongst them, however, there were some strangers that didn't combine readily with ordinary, everyday pro-

* BBC Radio broadcast, 14 November, 1976.

tons and neutrons, but seemed to keep themselves to themselves. What enigmatic quantity distinguishes these particles? What can they call such an attribute, with, it seems, no other handle by which we might grasp it physically? Forget your Greek and go to the international vernacular of modern science. From now on the English word *strangeness* shall be the name of what is to be added together and equated in such reactions. So now it is not just comical pedantry to say of a neutron or a proton that it has 'strangeness quantum number zero' to show how very ordinary and common-place it is! It is a scientific statement, pregnant with inner meaning.

Recently the same problem arose in connexion with the various families of quarks. Why are some particles much more stable than others? The latest experiments confirm the hypothesis that some quarks have a special property called *charm,* which must never be forgotten or changed. Indeed, the latest exciting discovery in high energy physics may be *charmonium* itself, an ideal partnership of a *charmed* quark and an *anti-charmed* quark, waltzing around one another in sublime perfection until, after what seems an inordinately long life – a charmed life, we might say – their mutual suicide pact carries them away in a flash of radiation. Oh shade of William Whewell; forgive us our barbarism in our scientific Tower of Babel!

What was that noise I made then? *Quark! Quark! Quark!* – like the cry of a seagull. So perhaps James Joyce meant when he began a new chapter of *Finnegans Wake:* 'Three quarks for Muster Mark'. Mark was evidently King Mark of *Tristan and Isolde,* on a voyage. But Joyce could not have been unaware that *quark* is the German word for soft cheese or, metaphorically, for rubbish. Murray Gell-Mann chose this word in 1964 for the hypothetical entities that go in threes to make a proton, giving no more than the page reference to *Finnigan's Wake.* The defining phrase is quoted widely, but I must admit that until I checked all the commentaries standing side-by-side in academic order on the shelf in the Univesity Library, I thought it must be a 'kwork', a measure of drink, something ancient and Irish, three of them to the gallon, a superior 'quart' for a superior boozer in a splendid Dublin tavern. It's good to know that those who hunted the *quark,* with linear accelerators and bubble chambers, need no longer fear that it is a cousin of the enigmatic *snark,* which was, of course, a Boojum, and never was seen again.

Of all the naming myths, I prefer the spontaneous response of Enrico Fermi, the great Italian physicist. In the 1930s there was talk of a new neutral particle to explain radioactive decay: could it be the *neutron,* recently discovered by Chadwick? 'No, no,' said Fermi. 'This is not Chadwick's neutron. It is a small neutron. It is a very, very small neutron. It is – it is –a *neutrino.*' And so it has been, to physics, ever since; bastard of Greek and an Italian diminutive, but alive and well, thank you very much, in our enigmatic Universe.

PART THREE

SCIENCE AS A PROFESSION

15

Scientists: Gentlemen or Players?*

Research has become a profession – although not quite so flourishing as when this was written.

Can you pay a man to think? 'In modern science the era of the primitive church is passing, and the era of the Bishop is upon us. Indeed the heads of great laboratories are very much like Bishops, with their association with the powerful in all walks of life, and the dangers they incur of the carnal sins of pride and lust for power.' That vivid simile is attributed to a great scholar, a Cardinal Archbishop among scientists, the late John von Neumann. It sharpens into a sword-blade one of the great questions that our society will soon have to face: can the fact that scientific research is now pursued in permanent institutions manned by highly trained professionals be reconciled with its being necessarily the vocation of the unusual individual?

Let us not doubt that scientific professionalism is already with us. Look at the Sunday newspapers with their pages of advertisements of jobs for physicists, chemists and engineers. Look at the terms of these appointments: 'Senior Principal Scientific Officer'; 'not less than £1800 per annum'; 'salary according to age and experience'; 'contributory superannuation scheme available'. What have these strange phrases got to do with our traditional image of a scientist – Darwin in his cramped cabin on the 'Beagle', or resting deep in thought on a couch in his country home; Newton isolated in the country by the plague, or hearing the clocks chime out over Trinity Great Court; Pasteur struggling with rabid dogs; Green, the miller; Lavoisier, the tax farmer; Dalton, the schoolmaster? These men were not paid to do research. They were great scientists because their minds were on fire, because they could not help themselves any more than Francis of Assisi could avoid being a saint, or the Brontës avoid writing novels.

Professor Stephen Toulmin, as befits a philosopher, sees this professionalism as a threat to the *intellectual* content of science. He fears a softening

* BBC Radio broadcast, published in *The Listener* 63,599–601 (1960).

of thought into orthodoxy and conservatism, and the loss of the high spirit of adventure and originality. This may be a danger, but I confess I do not see it. Indeed 'originality' has become the only scientific virtue, and careful execution of difficult work is scarcely recognized or rewarded. In my view the dangers to science are more subtle and yet more mundane, more personal and yet more social. They have to do with the sort of people that scientists are, the houses they live in, the salaries they enjoy, the education they get, the jobs they hold down, rather than with the sort of ideas they permit themselves to think. The symptoms are of a social, institutional malady, that could take the joy out of scientific work while still leaving it more or less capable of its official task.

For example, what sort of people become scientists nowadays? Surely not just those who are born to it, those like the Elephant's Child 'full of 'satiable curtiosity', those with a pathological inability to believe what they are told. The young men and women who are taking those highly paid, pensionable, graded posts are not social oddities but normal, capable, well-adjusted persons, setting up in respectable positions, hoping to make for themselves an honourable place in the world. It is a career open to the talents, like medicine, or the Church, or the Bar. Let me not suggest that they are governed by mere worldly ambition. A competent scientist with a doctor's degree can make perhaps £1500 a year in his thirties; for the same effort one would do much better as an accountant, or a solicitor, or a bookmaker. The struggle for the most esteemed posts is intense, and few get to the very top. No, they are borne along by the flood of romantic idealism about science that swept through our society in the wake of Mr H. G. Wells. Tens of thousands of schoolboys have no idea in their heads but that they too will split the atom or ride on a rocket round the Moon. Ask the average science student what he wants to do, and he will say 'Research' in the same tone, one fancies, as in the twelfth century he would have said 'Crusade'. As he sees it, science is an interesting and honourable profession, to be contrasted with what he believes to be the drabness of school teaching and municipal engineering, or the moral uncertainties of advertising and journalism.

All the same, enthusiasm and moral virtue are not enough; a certain modicum of intelligence is also required. Here is our sole criterion for entry into the profession: the ability to leap nimbly over all those hurdles – eleven-plus, O-level, A-level, college scholarship, triposes and degrees – that we put in the way of our aspiring mandarins. Formal intelligence is a necessary quality in a scientist, although there are interesting cases of distinguished scholars who did badly at examinations, and we should be most careful not to set up watertight barriers againt the exceptional man.

Unfortunately, skill at answering questions and solving problems is not the only intellectual virtue needed in research. There are other ingredients of temperament and intellect – tenacity, concentration, energy, imagination, insight, curiosity, and so on – which cannot be detected so easily, and which may only appear when a man actually settles down to a research problem. In theory all these are tested in those few years when he is a research student. He will show by the presentation of a thesis that he is capable of making a substantial contribution to knowledge, whereupon he is awarded the misleading title of Doctor of Philosophy, and released on the public.

See now the problem we face in our big graduate schools. Every young man with a good science degree claims the right to do research, and there is money to support him. With no way of testing their ability for research beforehand, we must take them all, and give them the training they demand. In principle, each student should be given facilities and a certain amount of inspiration from some great leader of research, and then be left to find and solve his own problems. This traditional method of learning to swim by jumping in and nearly drowning is certainly an excellent way of selecting the born scientists; but unfortunately not many of our masses of merely talented students could survive the ordeal by water. Moreover, even the best scholars can benefit from more formal instruction. The growth of knowledge requires each generation to 'stand on the shoulders of its predecessors, and see a little further', as Newton put it. We are training men now for a career of 40 years of science; they must have a broad foundation of knowledge to appreciate all the advances that will come in those 40 years.

Thus the years as a research student are a compromise, and a rigorous test of ability for research is never made. We set our students problems they can do, we help them with hints and advice, and, in the end, we 'get them through' their Ph.D.s. There is a tremendous demand for men with doctorates, with this professional certificate of competence, this master-mariner's ticket, this apprentice's diploma; who are we to deny a man a comfortable career if somebody is prepared to employ him? At best, our research student may have acquired something of a scientific conscience, a doubting demon always ready to see the errors in every measurement and the fallacies in every brilliant idea that springs into his mind. At worst, we may have taught him a technical trick or two that he can go on repeating like a nervous tic.

After the doctorate there are more years of effort and struggle before one becomes established in this highly competitive profession. Achievement now is measured in published work, so that our journals are bloated with the honey of 100 000 research-worker bees each trying desperately to make his name. Let me not digress upon this subject: the problem of finding out what has actually been published, the responsibilities of referees and editors, the irresponsibilities of some publishers, the incoherence of authors, the rubbish that gets accepted, the nonsense one has to read, the preprints and reprints and letters and private communications and references and review papers and research notes, and all the other technical business that goes with scientific publishing. Suffice to say that publication is essential to science; because it is also the main measure of professional standing it has become like pea soup, concealing the very morsels of truth that make it nourishing.

Suppose that, by the time he is 30, our clever young man has 'arrived', whether by a single brilliant discovery or by a series of able papers each making a distinct advance. He will then enter into the international community which is the modern equivalent of the guild of scholars that once united Christendom. He will fly round the world to attend conferences, he will be asked to write books and review articles, he will lecture here and there for a fancy fee, and even appear on television. There may be offers of glittering jobs, and a tussle over his body and brain between one university and another. This is perhaps the gravest peril of professionalism. In the United States the 'star' system is already well developed, and scientific integrity is seriously threatened by the corruption of limitless wealth and power in the gift of government and private industry. In this country we have the advantage of non-material rewards for ambition – college fellowships, professorial chairs, the Royal Society, knighthoods, and all the other trappings of status that our society so delicately provides. But we also know how to ruin a man for science by giving him a job with more 'responsibility', more men and money to command, more committees to bother with, more administrative decisions to clog the mind. Did we not make the incomparable Newton into a Top Civil Servant?

But if the bright lights of success or fame are only for a gifted few, there is work for everyone. There is no reservoir of disgruntled, underemployed professionals, as among actors and barristers, grimly hanging on, hoping for jobs. Those with the ability to solve practical problems reap rich rewards in applied science and technology. Others may be content to work in a team, to be the agents of more powerful minds. Gone are the days when, with sealing wax, string and a modest skill at glass blowing, one could make deep and devastating raids into the unknown.

The easily exploited regions are all invaded, and we can progress now only with complicated and expensive machines: cyclotrons, electron microscopes, Moon rockets, nuclear reactors, digital computers, and all the gadgets, services and technicians of a large laboratory. Big battalions are needed; even if every private carries a marshal's baton in his knapsack we cannot now be like South American armies, all generals and colonels. The graduate scientist, with his doctor's degree, ceases to be an independent artist, and shades off into the engineer, the technologist, the craftsman with skill and experience but a circumscribed realm of work. Taken in the proper spirit, these jobs are interesting and noble. There is the satisfaction of achieving one's own part, of having done good work, of not having let down the team. Yet this admirable spirit does not come easily to a man whose training was in a different mould; the whole mystique of the professional scientist is individualistic, and his main virtue is supposed to be his sturdy independence of spirit.

Then there are some who never really enjoy research, and who should get out of it. But professionalism breeds pride and snobbery, so that it is difficult to admit that one would be happier teaching school, selling radio sets, or making money. With our M.Sc.s and our B.A.s and our Ph.D.s we are labelled as dedicated, and it would be almost as defrocked priests that we would go out into the wide world and take ordinary jobs. What was once a free activity has become a caste.

Thus there is a serious tension between professionalism in science and the free, amateur spirit that must be in the heart of every scientist. There is a contradiction in the idea of deliberately selecting students for training as scientists. There is a discrepancy between training in research and original research itself. There is an antithesis between ambition for professional success and the disinterested search for truth. There is antipathy between the team spirit and the freedom of the individual to follow his bent.

Some of these tensions are inevitable, and none of them is fatal to scientific knowledge, which has an astonishing power of survival in adverse circumstances. But we must be realistic about the problems and learn to live with them. We must, to some extent, tame the wild romanticism of our students, and let them see the scientific life as it really is. We must select and train them carefully, neither spoon-feeding the able nor encouraging the pretensions of the weak. We must not give pure-research jobs to those who lack the proper inner sources of intellectual energy. We must take routine tasks off the minds of the best researchers, setting them free for their true vocation. We must curb the corruptions of wealth and power, that can make scholars into the performing apes and courtly fools of princely governments and corporations. Above all, we must remember that scientific inquiry can never be a job, to be performed at piece rates or by the hour; it is a free activity of the human mind, fascinating, dangerous, and exciting, to be done for its own sake. It is a drug for which some men have no taste, yet which is, for a few, the food and drink of life itself.

16

A Very Strange Tribe*

To explain to a layman the sociological significance of the activities of High Energy Physicists ('heps') and High Energy Theorists ('hets') one is forced to use an anthropological analogy.

This book about my neighbours, the *uk-hep* and *uk-het* clans, contains much interesting information that will be used by other anthropologists in their analysis of our peculiar culture. Yet there are many aspects of clan life and work that would not be immediately apparent to a foreigner who had not been brought up in our native ways. This is particularly difficult in the case of the hep and het clans, whose rituals are performed to the accompaniment of songs in a cryptic symbolic language whose meaning cannot be translated to the uninitiated. As an *uk-solstath,* however, I have had the good fortune to learn some of these songs for use in the rituals of my own clan, and have also had many opportunities to observe the uk-heps and uk-hets in their daily lives. I therefore felt it my duty to the science of anthropology, in commending this book to you, to sketch in some of this background as seen by one of the natives.

The first point to make clear is that all members of these clans have dedicated their lives, by a series of magical ceremonies of great psychological weight, to a single purpose – the cultivation of the *elpar* fruit (*Nucleonicus fermii*). Whatever secular power or worldly wealth may accrue to them from public success in this subtle and highly competitive art, is almost irrelevant besides the fascination of horticulture itself. Those who have been called away into the very highest ranks of chieftainship in the tribe as a whole, whether in warlike or peaceful pursuits, always speak with genuine regret of the laborious days and nights they once spent tending the young plants, controlling the irrigation channels, examining the harvest of husks, and selecting the sweetest-smelling kernels. Despite the unworthy spirit of competition that occasionally seizes upon individuals and groups when some new variety is coming to seed or a new crop

* Foreword to *Originality and Competition in Science: a study of the British High Energy Community* by Jerry Gaston (University of Chicago Press, 1973). See p. 98 for a glossary.

is ready for picking, the normal attitude is one of friendliness and cooperation between all members of the clans. The social psychology of clan life puts great emphasis on such fraternity, to the detriment of good relations between members of different clans. All heps and hets of whatever tribe firmly believe that the cultivation of the elpar (which is, incidentally, quite inedible) is the most important thing in life, and that all the tribal resources should be devoted to this single end. Personal rivalries within a clan are quite insignificant by comparison with this fiercely sectarian attitude. This clannishness even extends across tribal boundaries. As we shall see in a moment, the hep and het clans are so distinct in their social roles that one may find closer fraternity between, say, uk-hets, us-hets, eur-hets and russ-hets, even though these belong to different tribes that are often at war with one another, than between the uk-hets and the uk-heps alongside whom they actually live.

The differentiation of role between heps and hets arises quite naturally out of the peculiar life cycle of the elpar. The reproductive phase, in which the seeds germinate rapidly, grow to young plants, flower, and produce a few new seeds, is under the charge of the hets; the fruiting phase, which lasts much longer, and leads to a large crop, is in the care of the heps. The horticultural activities appropriate to these two phases are so entirely different that they lead to entirely different styles of social organization within the two clans. Between solstaths and solstaphs we also have a similar differentiation, but the distinction between the two clans is not nearly so marked as it is between hets and heps, who stand at almost opposite poles in the horticultural temperament.

The sole aim of the het is to breed a significantly new variety of elpar, as measured by the beauty of its flower and eventually by the fragrance of its fruit. This is essentially a solitary task, demanding no more than a few seed trays and pots, a little fertile compost and a modest supply of water, light and air. The secret of success is thought to lie mainly in the choice of ritual chants, which must be sung with perfect accuracy and clarity as the seed germinates and as the young plant pushes out its leaves and flower buds. The hets are, incidentally, the supreme masters of the symbolic language, and often vie with one another in the composition of more and more sophisticated and esoteric songs. But the individual het usually works alone, or in voluntary collaboration with two or three of his fellows, thus avoiding the dangers of mistakes arising from mutual misunderstandings as to the text actually being chanted. Such mistakes are especially prevalent when the attempt is made to breed new plants on a large scale, using numerous junior assistants to sing standardized songs which they do not fully understand. Hets cooperate mainly in the exchange of seeds, and of the appropriate songs, or in occasional sessions of mutual fertilization of their flowering plants, for the elpar quickly becomes self-sterile after a few generations of in-breeding. Such exchanges are indeed so essential that many young hets, lacking confidence in their own stocks of seed, form cliques of 'clients' around their more notable and experienced colleagues, raising minor variants of any new variety that may thus be handed them. But such informal social groupings are essentially voluntary, and a het patron has no enforceable authority over his clients.

The heps, on the other hand, are tightly organized in distinct groups. The members of each group work together as a gang, under the direction of an experienced headman. The harvesting of a satisfactory crop of elpar is extremely expensive and laborious, calling for great communal resources and efficient planning of specialized tasks. The enormous irrigation works that supply water continuously over a large area of fertile land are now no longer the sole responsibility of the hep clan that originally planned and built them, and are shared by many such gangs, but much labour is still required to prepare the land for each new crop. The young plants – usually acquired from friendly hets – must be set out according to a strict ritual pattern, and watered, weeded, pruned, etc. over a period of several years until the fruit has formed. But only a small proportion of the vast crop of husks actually contains a ripe kernel. One of the most exacting duties of a member of a hep gang is to oversee the work of several scores of slave women who break open each husk and look for the tiny kernels. These, in turn, must be closely scrutinized for signs of decay or other blemishes before they can be added to the accumulated harvest. At every stage, therefore, the work is arduous, time-consuming and highly specialized. Each member of the gang must know his own job, and carry it out with perfect precision in coordination with his fellows, if the whole crop is not to be a failure. The responsibilities hang heavily on the headman, especially since he must, in his turn, subordinate the work of his gang to the overall irrigation programme of the tribe.

The social organization of the hep clan is thus naturally hierarchical, with authority in the hands of the caste of headmen. Rivalry between individuals within a gang is firmly suppressed and competition between gangs in the same tribal region must be controlled. The competitive incidents reported by anthropologists have almost always involved hep gangs from different tribes – uk-heps in competition with us-heps, for example. Yet even these tensions are repressed at the great corroborees of hets and heps, where seeds and plants, ritual chants and fruit, are compared, judged and exchanged in a great orgy of friendship and mutual congratulation. Anthropological studies that concentrate on the occasional breakdown of these norms miss the most significant features of our clan system.

The most subtle aspect of this system is the symbiotic relationship between the hets and the heps. At first sight one would suppose that the heps are clearly subordinate, their task being simply to bring into production each newly-bred variety. But the truth is more complex. It is often argued, for example, that it is the hets who have the more lowly role; theirs is the effeminate task of growing a few seed plants for the manly, physically active, hep gangs which do the real work. The most thoughtful hets themselves emphasize that the quality of a new variety cannot be judged solely by the beauty of the flowers, but must eventually be tested by the fragrance and quantity of the fruit. In some cases, such testing takes many years – too long a time for those impatient hets who practice a programme of random crossing without selection, hoping that a valuable new variety may occur by chance amongst the many seeds they produce.

But these subtleties are well understood by the leading members of the clans, and the award of high status amongst uk-hets by no means goes to those claiming to have produced the most new varieties of seed. In particular the rank of Fel-roy – a title of nobility that completely outweighs all other measures of social standing among the clans of uk – is given to an uk-het only for the breeding of some outstanding new variety that has been thoroughly tested in cultivation. The fact that this rank (for which there is no real equivalent amongst the us-hets and us-heps) is occasionally won at an early age shows how much more weight is given to quality than to quantity of production.

Amongst the uk-heps, although some of the leading headmen are Fel-roys, effective rank is much more closely linked to authority over a gang. A headman with a gang of sturdy heps working in a high flow irrigation system can acquire great standing in the clan, not so much by his own personal achievements as by the labours of his gang. But the assignment of public rank amongst the members of such a closely organized group presents many obstacles, owing to the specialized division of labour and the long time needed to harvest each crop. In practice, therefore, the judgement of the headman concerning the ability of each member of his gang is paramount. This is one of the key features of the hep clan as a social institution – a characteristic that contrasts strongly with the artistic individualism favoured by the hets.

Our clans are not, of course, hereditary in the strict sense but are recruited by adoption. In the land of uk it is universally believed that the trait of *wrangler* – loosely translated 'green fingers' – although rare amongst the people as a whole, may infallibly be detected at an early age by magical incantations in the ritual language. Young men and women in whom this highly-prized quality has been observed are then given special horticultural training by which their natural ability is supposedly strengthened. A further ceremonial test in their twenty-second year decides their fate. A few of the most superior wranglers are then adopted by one or other of the most experienced hets, since green fingers are obviously much more valuable in the raising of delicate new varieties than in the large-scale production of a crop. The next grades of initiates are adopted into the hep clan – and retain throughout their lives a lingering jealousy of the hets whom they once aspired to emulate. But the training of heps is so specialized and toilsome that they are seldom able to learn the more powerful ritual songs and are soon cut off from the skills of seed raising. This explains both the division of labour between the clans, and, to some extent, the subtle feelings of superiority and inferiority between them.

But to understand the system in all its dimensions one must go beyond the static picture presented by many contemporary anthropologists, who seem to suppose that the same social institutions have existed from time immemorial. The uk-het and uk-hep clans date back to the beginning of this century, to the great headman JoJo and his adopted son Ŭth, who was himself the headman or adopted father of many of the older hep headmen of our day. Ŭth himself was what we should now call a hep, but he had an unusual knack for raising many sturdy new varieties of plant without the aid of ritual songs, and was indeed somewhat scornful of many of the more fanciful varieties being bred by the hets of his day. This attitude also, transmitted via his adopted sons, is by no means outmoded amongst the present generation of uk-heps, who still maintain amongst themselves something of the family feeling of the founder of their clan.

I must apologize to readers for this non-scientific description of the psychological and historical aspects of our tribal life, but it seemed that many of the observations recorded in these anthropological researches would lose their significance unless seen in this context. As a humble member of a neighbouring clan who has had the good fortune to become acquainted with several visiting anthropologists, I felt it might be of some use to say some of these things in my own simple fashion.

Glossary

uk-het:	United Kingdom High Energy Theorist
us-hep:	United States High Energy (Experimental) Physicist
elpar:	elementary particle
solstath:	Solid State theorist
solstaph:	Solid State physicist
irrigation works:	particle accelerator
Fel-roy:	Fellow of the Royal Society
JoJo:	Sir J. J. Thomson
Rŭth:	Lord Rutherford

17

Spotting Winners*

Should successful research be its own reward?

How should we think of scientists as a community? Some sociologists have been telling us that we are really *Trobriand Islanders,* exchanging gifts of publications for recognition. The *Mafia* model is adopted by certain journalists raking muck from amongst the very pillars of the scientific Establishment. In some quarters, the categories of the *Class War* are fitted to the professoriate and their exploited assistants. It is certainly a relief to read a book whose starting point is the simple assumption that most scientists try to follow the accepted norms of universalism, humility, disinterestedness, etc., and regard as legitimate the authority and power exercised by those who are made famous for their personal contributions to scientific knowledge. Despite all one's suspicions to the contrary, it is not inconceivable that the scientific community approximates roughly to a Utopian Republic, in which true merit is rewarded and the government resides with the brightest and the best.

The strategy of this book is to test this hypothesis against various alternative propositions. The tactics are straightforward. From the *Science Citation Index* one can determine the number of times that any particular scientific paper has been cited by other scientists in subsequent years. This is only a crude measure of the quality of a paper, but is probably the best way of producing an objective index of the scientific contributions of its author. This index can then be correlated statistically with various other measures of 'authority' or 'achievement' such as prizes, employment at a prestigious institution, 'visibility' to other scientists, etc. The study is confined to American physics, but could easily be extended to other disciplines and other countries.

The fundamental conclusion is (alas?) that our suspicions are unjustified. The evidence is that the prestige accorded to the leading physicists

* Review of *Social Stratification in Science* by J. R. Cole and S. Cole (University of Chicago Press, 1973), published in *Science Forum,* June 1974.

is pretty much what they deserve, and that they do not seriously abuse their privileges. Various hypotheses alleging conspiracy, favouritism, nepotism, snobbery, and other vices are falsified by the facts. Even the cynical doctrine of 'publish or perish' turns out to be invalid: those who emit large numbers of papers that are seldom cited by other physicists get nowhere by their efforts. Evidence for discrimination against women, beyond the Ph.D. hurdle, is practically zero; they achieve the scientific standing, the prestige, the jobs, that their scientific publications earn them. As the authors themselves sum up:

> If the social system of science operated on completely rational and universalistic principles, quality of role-performance would be the sole criterion upon which scientists would be evaluated, and all scientists producing high-quality work would be rewarded, regardless of their other characteristics. Their non-scientific statuses, such as age, sex, race, and religion; their scientific origins (where they earned their doctorate) and their location in the scientific social structure would all have no influence on the amount of recognition received . . . The general conclusion of our research has been that science does to a great extent approximate to its ideal of universalism.

To understand this, we must go a little deeper than mere conformity to ideal norms of scientific behaviour. The implication of the citation indicator is that every scientist lives intellectually in the milieu created by his predecessors and contemporaries. He simply *must* know what they know, and learn to think as they think. His task is not (as rather suggested here) to *maintain* a consensus but to *create* one, by unearthing new and persuasive evidence, by well phrased criticism, by acceptance of the necessary, by postulating new interpretations. The active scientist must be sensitive to all relevant messages that may come to him, as he continually fashions and refashions his own image of Nature and tries to impose it on his colleagues.

But these messages are immensely variable in significance. What this book brings out is the desperately skew distribution of ability in research. This must be measured on the Landau scale: 'The first-class scientist achieves ten times as much (in quality and/or quantity) as the second-class', and so on. To pick up that rare first-class message, or to choose that rare first-class man as his student, assistant, or colleague, every scientist must cultivate a very sharp ear for genuine quality. This sort of judgement is not a mere game; it engages him right to the heart of his professional skill. In science there is no independent role like that of the art critic; to do good research at all, you must be able to assess the research of other people.

The real question is the extent to which it is functionally valuable to reward good work. The Professors Cole miss an important point by failing to discuss the most functional form of intellectual reward – the invitation to write a definitive review article or to give an 'invited paper' at a scientific conference. More directly offered, and yet more humbly taken, the compliment is paid: 'Please tell us, first, what you think about it, before we try to go any further.' This is the actual role demanded of the 'authority' – not to reaffirm a dead consensus, but to bring life to a new one. This is the sole level to which the ideal scientist should aspire – to voice to a meeting of his own international Invisible College his view of their common enterprise and its current state. Their approval and applause is 'recognition' enough.

Do we really need an overt system of awards – Nobel prizes, memberships of National Academies, etc. – above this level? Are the actual distinctions thus made somewhat too invidious, creating artificial strata in a many-dimensional continuum of talents and styles? One of the interesting discoveries recorded in this book is that recipients of prestigious awards are much more aware of the existence of other similar awards than their lowlier brethren. Is it not possible that we simply breed in them a vanity that feeds upon its own gratification; like noblemen at the Court of France, becoming a little too sensitive to precedence and trivial tokens of honour? Much that is reported here concerning the 'Matthew Effect' – 'to him who hath shall be given' – applies with real force to the accumulation of miscellaneous prizes by star scientists.

What we should all be concerned about is that intellectual authority should not be extended outside its proper sphere. The society whose stratification is discussed in this book is essentially that of the Invisible College, more vivid in the imagination than in actuality. The Great White Chiefs of physics do, indeed, dispose of large resources for research, but they do not fully govern the professional lives of the scientists to whom their committees award contracts. An academic scientist owes allegiance to his own university, to his department, to his students, as well as to the national leaders of his special discipline. There are other bases of power in the scholarly and technological communities which are not legitimated by reference to the *Science Citation Index* or to the whims of the Swedish Academy. As mediators between the scientific community and society at large, we must recognize not only powerful administrators and civil servants, but also the popularizers, expert consultants and other pundits, who exert a great deal of influence from quite slender foundations of research talent.

From the evidence presented in this book, it would be easy to suppose that scientific meritocracy is so rational that it ought to be applied systematically to all aspects of the scientific life. On the contrary, the result would be disastrous. The system works now because, as the authors point out, failure in research does not carry heavy material penalties. Things have changed a bit in the last few years, but whilst science and higher education were expanding rapidly the man with a Ph.D. never really had to fear being out of a job. The impression that there is no 'class conflict' in American physics stems as much from the existence of this expanding frontier of opportunity as from the idealistic norms of science. The situation now in France, for example, with new appointments and promotions in the CNRS almost completely blocked, looks much less rosy.

What we must recognize is that for the leading academic scientists, research is still a vocation rather than a profession. However much they are involved with big machines, large teams and other bureaucratic lumber, they see themselves as independent scholars and teachers, not as functionaries. They have always been able to say to themselves that they might soon give up research and devote themselves to teaching, administration, academic politics, and writing a book. Failure, or the success of others, has little leverage on such people. This is the true writ of 'academic freedom', which still runs in American physics. But replace this with a more rational functional system of full-time professional research, organized bureaucratically and hierarchically according to the citation indicator and you create a monster. It is one thing to fade gracefully from research, complaining of the burden of teaching and administration; it is quite different to stay for years as a junior assistant researcher, or to dismissed because nobody has yet cited your Ph.D. paper!

The Coles do not say quite that; but they do argue, quite correctly and cogently, on the basis of chains of citations, that science is done mainly by its own élite, and owes little to the contributions of Landau's fourth and fifth classes. The Matthew policy of favouring the most promising is certainly wise. But to go further, as they suggest, and deliberately chop off the bottom half of the physics community might be counter-productive in more subtle ways than they envisage. It depends on how quickly, and by what machinery, the trick is played. A Ph.D. without published papers to his name is a tolerable misfortune, but a Ph.D. without a job is a human tragedy. Who profit most from a Mafia: the rich or the poor?

18

Some Very Queer Fish*

Social factors do not entirely provide the motive force of the scientific mind, in the seventeenth century and now.

The first title rings a bell: of course – this is a reprint of Professor Merton's doctoral dissertation, as it appeared in *Osiris,* in 1938, to which the author has added a new preface of more than 20 pages. This classic of the sociology of science has had a very considerable influence, not only on our view of seventeenth-century science but also as a pioneering effort to study the dynamic relationship between society and science. In his 1970 preface, Professor Merton is modest about his youthful work; even from his present eminence as the doyen of this subject, he need not deny that it was a formidable scholarly achievement.

It is best known, of course, for its advocacy of the hypothesis that the flowering of science in seventeenth-century England was a product of the Puritan movement of religious reform. This idea comes from Max Weber, as a corollary to the more familiar causal relation between Protestantism and capitalism, but Merton argues it out in detail, and makes it his own, both by the factual evidence that he piles up and by his theoretical interpretation of the intellectual and social forces at work. He also gives a great deal of attention to the intimate connexion between pure science and material techniques, showing the importance of mining, navigation and other useful arts in the literature of the time.

The basic evidence on the Puritan connection is pretty convincing (despite Feuer's lengthy counter-arguments) in that he proves that scientists, then and since, have been much more Protestant, Jewish or agnostic than they have ever been Catholic in their acknowledged beliefs. This is surely an historical fact. What is not quite so sure, and cannot be proved conclusively by the available evidence, is that modern science arose as a *positive* creation of Puritan theology, metaphysics or ethics. For my taste – and perhaps now for his own more mature judgement – Merton argued this a

* Review of *Science, Technology and Society in Seventeenth Century England* by R. K. Merton (Harper & Row, 1971), and *Scientists at Work* edited by T. Dalenius, G. Karlsson and S. Malmquist (Stockholm: Almquist & Wiksell, 1970), published in *Minerva* 9, 434–7 (1971).

little too closely, squeezing every last drop in favour of his position from the published literature of the Puritan movement. On paper he is wisely balanced and undogmatic in his interpretation, being well aware of the manifold social forces at work at any time in any society, but the effect of this book is undoubtedly towards convincing us that this particular 'idealist' influence was dominant within the complex of causes.

Without the historical scholarship to back me up, I can hardly offer detailed opposition to this thesis, but I cannot help feeling that it is somewhat too strongly phrased. Would it not be better to see the rise of science in that particular era as a consequence of the Renaissance/Reformation revolt *against* the doctrine, based on revelation, of the medieval church. The scientists of Protestant England were not so much pro-Puritan as anti-Catholic; it was the overthrow of *dogma,* as promulgated by the Bishop of Rome, which released the hormones that stimulated the flowering. Science was something new and 'progressive'; was it likely to appeal to those who by taste adhered to 'the old religion'? Have we not seen, in our own time, a strongly Marxist tint in the religious beliefs of many leading scientists, not so much in response to the claim of 'scientific socialism' but as an emotional prejudice in favour of new and revolutionary modes of thought?

Even this would be too simple, for modern science is as much the child of medieval scholasticism, with its search for logical precision and a consensual world picture, as it is of bourgeois empiricism. Merton seemed to care nothing for the achievements of the Middle Ages in mathematics as well as in productive techniques. He almost ran into the characteristic sociological error, of neglecting the historical background of the social system he was studying. For such *Zeitgeist*-like phenomena as he was considering, a mere century is by no means a sufficient span of time.

One begins to feel, indeed, that the sort of ideological sociology which was fashionable in the 1930s, and in which he embedded his argument, was altogether too schematic and mechanical. By taking various idealizations and rationalizations from sermons and prospectuses, one does not altogether explain how and why people do what they do. There is an implicit social psychology of simple norms and characteristic influences on average people in this interpretation, which is fair enough for the analysis of voting behaviour in a two-party democracy, but which is not really consistent with our knowledge of the sheer oddity and marginality of the scientist in those happy far-off days.

This is beautifully demonstrated in *Scientists at Work,* which is both a *Festschrift* for a distinguished mathematical statistician, Herman Wold, and a collection of essays on scientific creativity, motivation and methodology. A number of first-rate mathematicians and social scientists were asked by the editors simply to describe the way in which they work at research and how they get ideas. It was all said long ago in a famous essay by Poincaré; but it is refreshing to be reminded that eccentricity and anarchy, serendipity and obsession, counter-suggestion, jealousy, paranoiac suspicion, spasmodic laziness, arrogant virtuosity and other individualistic traits are still to be regarded as essential ingredients in scientific creation. Some of the authors could only work when quite alone; others when in company; some need to be unhappy; others prefer serenity; some are spurred on by the desire to do the other man down; others are motivated by pure curiosity.

Mathematicians, of course, are lone birds in their actual thoughts, and do not do 'big science' in organized teams. They cannot fill up their days with industrious make-work such as resoldering bad connexions or debugging their computer programs. But they present an authentic picture of the real scientist – the dilettante who becomes obsessed with his little problems until he nearly gets run over as he stops in thought in the middle of the street. It is good to know that the poor old soul has not been made redundant by a lot of smooth jokers with a soft professional patter, treatises on efficient research management and portfolios of big fat contracts. I would recommend this book to any aspiring scholar-apprentice, especially for Galtung's beautiful essay and for Naess's concluding sentence: 'As long as we do not pretend to be competent, no harm is done.'

The relevance of this book to Merton's thesis is that surely the old heroes of the seventeenth century were just like this. In any human population, there are always to be found such suspicious, introverted, curious, pedantic, obsessive, argumentative characters as our old-fashioned scholars. The sociological question is how they get to be doing science, instead of just going to law against their neighbours, devising curious toys and games for their grandchildren, or sitting under the peepul tree spinning fantastic yarns about how the world began. Much more than a positive ideology of science, they needed literacy and leisure. In the Middle Ages these privileges were available only to churchmen and they spent their speculative talents in the tangles of scholasticism. The Reformation did not make experimental science respectable, but at least it was freed from taints and heresy. It also created a substantial class of people with the financial means and the basic education within which these particular obsessions could be indulged. In other words, we could see the scientific revolution of the seventeenth century, not as the triumph of a new general world-view, but as a social fluctuation in which these marginal personalities were enabled to develop their peculiar talents without gross repression.

They also began to find each other. Think what it must have meant to that queer fish Kepler to discover a fellow-spirit in Brahe or Galileo. Imagine yourself to have seen a genuine flying saucer in your backyard, yet reluctant to speak about it for fear of suffering ridicule or worse; then you meet another lunatic who hints that he has seen it too. The importance of those little scientific societies which sprang up in the seventeenth century was not their public face but the mutual support and encouragement they gave to their members, the opportunity to meet, discuss, communicate and criticize. In his final chapter Merton makes this point – but it could have been more central to his whole theme, and provided quite as efficient a prime mover for the social machinery of the scientific enterprise as any theological doctrine.

And of course they had a go at practical problems as well as the more solemn theoretical questions about the Earth and the Moon. It is only our own intellectual sophistication that separates the attributes of 'Nature' – those events that exist independently of 'man' – from those which he himself contrives and on which he is dependent. Remember that they were enthusiastic amateurs, not yet laden with specialized scholarly roles, or academic restrictive practices, or cerebral gentility. Having seen some of the successes of rational analysis and new techniques in practical matters, they began to believe that anything and everything was possible, just as Fred Hoyle believes that politics and economics can be mastered as easily as astronomy – or, as Lord Blackett showed, that submarines could be discovered as easily as neutrons.

But we must not be misled by their *claims* of utility. Of course they wanted to impress people with their value to society, and to get royal approval for their activities; what better protection from charges of magic and mystery-mongering than a few noble lords on the board. The fact is, however, as Merton makes clear, that very few of them were really interested in making inventions in the practical Yankee manner. For example, the whole business of finding longitude at sea may be subtly misleading; it was not that this was the supreme practical need of the day, but that the progress of astronomy and mechanics had brought this problem to the surface, as a challenge to the best intellects of the day in the most advanced field of research. Any competent astronomer, by the end of the seventeenth century, could see that it was soluble in principle; the Newtonian paradigms were well established, and the race was on, within normal science, to devise a technique that was sufficiently easy and accurate in practice.

These sceptical remarks, untutored by genuine historical knowledge, are not meant as serious criticisms of Merton's theme; but perhaps they indicate the directions in which the sociology of science has progressed since his monograph was written. The fact that he himself has led many of these advances, especially into the study of the internal sociology of the scientific community, shows that his starting point was well chosen, and worthy of a visit even now. But I cannot help thinking that Swift's satirical caricature of the Academy of Lagado, with its projects for extracting sunbeams out of cucumbers and calcining ice into gunpowder, more nearly represented popular opinion of the Royal Society and its Fellows than did the rhetorical apologia of Bishop Sprat and other Puritan divines.

19

Some Pathologies of the Scientific Life*

Research is by no means a gentle art, free from psychological tensions.

Nowadays many students opt for a career of 'research' without any idea of the history or present state of the profession they are entering. By the accidents of specialized education they are depressingly ignorant of the most elementary facts concerning the most famous figures and episodes even within their own chosen disciplines. One encounters chemistry students who do not know that Lavoisier was guillotined, physicists who have never heard of Rumford and his cannon, microbiologists who know nothing of Leeuwenhoek – and so on. They have no idéa of the way in which modern science is organized and financed, nor any notion of the various forms of employment or office that may be open to them. These intelligent but uneducated young men and women gradually acquire the skills of research, and learn to practise their profession purely by apprenticeship to older scientists. They copy and conform to the conventions of the scientific community without any clear idea of what it is all about.

The bulk of the literature available to them is profoundly misleading. There is a familiar hagiography of our Estate – historical and biographical works praising the lives and works of the great scientists of the past and present. The well-read schoolboy models himself on some imaginary hero – Darwin, the 'Great Naturalist', say, or Einstein, the 'Intrepid Thinker' – who bears little resemblance to the real man. Of course, one should be buoyed up in youth by fairy tales of courage and persistence, where even the bitterness of rejection and neglect is turned into a sort of glory by eventual triumph – but such edifying stereotypes are an inadequate preparation for real adult life.

Searching for more serious information, we encounter those topics gathered under the general heading of the History and Philosophy of Science. These do, indeed, constitute a fascinating body of knowledge but

* From Presidential Address to Section X of the British Association, September 4, 1970, published in *Advancement of Science* 27, 1–10 (1971).

somehow they have never quite lived up to the claim that they should be taught to every candidate for a B.Sc. They do not seem relevant to the actual performance of research in any particular field; they seem more significant as history and philosophy than as science; in the end, what was once alive has dried up, and crumbles at the touch.

The missing, vital element is the social context in which we act. To understand what it means to be a scientist, one must see oneself as a member of a specialized group of people, engaged in a corporate activity with communal means and ends.

When I became interested in this subject, some ten years ago, I could say, as if coining a phrase, 'No, it's not the history or philosophy of science that's important; it's the *sociology* we ought to look at' – and might then have to explain further that I did not mean merely the function of science in society at large. Nowadays, the term 'Sociology of Science' denotes a professional sub-discipline, with its own conferences, reprint exchanges, consultancies, pundits and the other paraphernalia of an Invisible College.

I must confess, however, that this tide of professional sociology does not always appear to carry us nearer to the inner soul of science itself. The limitations of scientific sociology as a complete representation scheme for social life are well enough known; these limitations apply equally to the attempt to characterize a special community such as that of science. Until our scholarly machinery of observation and theory has been much improved, there is quite a lot to learn from 'informal' comments embodying familiar experience, significant anecdotes and the ordinary common sense of daily life. As a part-time, amateur dilettante, I claim the right to talk a little nonsense of this kind, which would not be permitted at all if I had ever earned a Ph.D. in the subject.

Now the problem is this: what are the guiding principles, the behavioural norms of the scientific community? What function do they serve, and how does the individual learn to abide by them? To put it another way: when one becomes a scientist, what intellectual, moral, or material constraints shape our actions and decisions? A list of such norms is now one of the standard items in the Sociology of Science.

But the very words 'norm' and 'constraint' suggest antagonistic socio-psychological forces of which, in healthy conditions, we may be only dimly conscious. To discover the boundaries of the territory over which these forces have sway, and to feel their full power, we need to move away from typical conditions of equilibrium and harmony and look for evidence of disharmony, conflict, and dysfunction – in other words, the pathological cases. To quote an essay on this very subject by L. S. Kubie:[1] 'The wider and more easily recognized deviations of pathology illuminate the "normal" for us, and sensitize us to slighter anomalies which we otherwise would overlook.' The same strategy is at the heart of one of the most famous works in the whole of sociology: Durkheim's masterpiece, *Suicide*.

A collection of pathological cases must inevitably seem like an attack on the whole body of science. In the present irrational mood of our intelligentsia, it would seem folly to add more fuel to the fires on which they are mentally roasting us – with our books piled high about us and our instruments stuffed down our gullets. Let me say, then, most emphatically, that I do *not* believe that the internal state of the scientific community is desperately unhealthy. Some of the phenomena to be discussed are mildly scandalous, but they are mostly rare exceptions that 'prove' the rules. Anyway, science is big enough and strong enough for us to admit some of its defects, even at the risk of being deliberately misquoted.

I want to discuss those occasions when the *internal* social and psychological mechanisms of scientific research break down. As I have argued elsewhere,[2] science is a communal activity, directed towards the creation of a rational intellectual consensus. Scientists are in tacit cooperation when they communicate their results to one another, criticize, recognize, refer to, appoint or promote each other. They act in the expectation that their contemporaries will behave according to certain conventions. Any serious breach of these conventions is a pathological symptom, deserving our attention.

The typical psychopathology of ordinary society is crime. Strangely enough, deliberate, conscious, fraud is extremely rare in the world of academic science. One would have thought that the material rewards of research were tempting enough – cosy university appointments, research grants galore, plenty of conspicuous travel – to encourage a noticeable amount of scholarly dishonesty. Yet the only well-known case is 'Piltdown Man', which is more of a monument to the absolute trust that we have in a reputable fellow scientist than an example of a grandly-conceived crime. I suppose there must be a fair bit of lazy cheating – experimental data filled in by Guess and by God, or theories stolen from little-known works and plagiarized – but I have never come across it myself. It made a good plot for a novel by C. P. Snow – they say that *The Search*[3] is essentially autobiographical – but such episodes are not at all typical.

One must not get the idea that scientists are all that admirable in everyday life. As I have already remarked, they will intrigue for political ends like any Jesuit, and can be as lordly as any consultant physician in the control of their juniors. They can deceive their wives, fiddle their tax returns, drive drunkenly, live beyond their means, feed parking meters, beat their children, and otherwise behave as antisocially as anyone else when the occasion demands. There are scoundrels amongst scientists, just as there are amongst lawyers, or architects, or naval officers. But I have a very strong impression that these traits are repressed within research itself.

Perhaps the rewards are too meagre, and the risks of being found out are too great. The scientist seldom works directly for money or other material means that give immediate gratification; he seeks 'recognition', or 'esteem', which are intangible and come slowly. No doubt one could easily devise a succession of modest frauds, in some relatively obscure field that would earn one a decent life tenure at a university before they were uncovered. But it would take a real genius to invent a bogus discovery, worthy of a Nobel Prize, that would not be discredited within a year or so. Banks, mail trains and the ever-gullible readers of company prospectuses are more fruitful targets for such abilities.

It may be, also, that the long training of the professional scientist conditions him against conscious deceit. Those years leading up to the Ph.D., the struggle to obtain useful results, to master the theory, to write it all down, and to please his supervisor, are a tough discipline. If he has studied in a good institution, then he will have internalized very high standards of honesty, scepticism and criticism, so that he will never find it easy to let his mind slide over the difficulties and objections, and thus to rationalize a really crude deception. He should have learnt also how completely each scientist relies upon the sincerity and good faith of those whose results he quotes as evidence for his own conclusions. Unwitting error causes trouble enough; the suspicion that any paper we refer to might be fraudulent would completely paralyse further research. A fierce and uncompromising honesty is one of the standard attributes of the so-called 'scientific attitude'. I do not think it is inborn, but it is certainly moulded into one by powerful social pressures in the graduate student phase of one's career.

Self-deception, on the other hand, is an exceedingly common phenomenon in the scientific world. We are all, of course, familiar with extreme cults, such as Flying Saucers and Extra-Sensory Perception, on the fringes of genuine science; it is important to realize that the reputable scientific literature abounds with similar, if less obviously fantastic, irrationalities. Some of the most famous cases – Blondlot's 'N-rays', the 'Davis–Barnes Effect', 'Mitogenetic Rays', the 'Allison Effect' – have been wittily described by Irving Langmuir in a talk entitled 'Pathological Science'.[4] The symptoms that Langmuir catalogues – barely detectable effects, claims of great accuracy, fantastic theories contrary to experience, criticisms met by *ad hoc* excuses – are surely significant for any philosophy or psychology of invention, and should be given close attention by any serious epistemologist.

What does this type of pathology signify for the scientific life and the scientific community? It is not surprising, perhaps, that the initial claim to some extraordinary discovery should get published. It is the duty of every journal editor and referee to have an open mind towards completely new effects, and to suspend his scepticism temporarily in order to entertain the possibility that some highly speculative and incomprehensible theory may, after all, be correct. But how are we to explain the very large number of papers that often then appear, both in support and refutation of some such preposterous proposal? The truth is, perhaps, that scientific progress is difficult and uncertain. It is well known that some of the greatest scientific discoveries have seemed bizarre and 'agin Nature' when first presented. The hope of participating in an important scientific revolution encourages many able professionals to follow up the initial claim, so that a 'fashion' may set in, with quite a body of optimistic and uncritical supporters to give it impetus. A considerable amount of experiment to establish negative results is then required, to debunk the effect or theory. The journals must be kept open to such communications for a year or two, and cannot easily refuse to publish the work of the master and his remaining disciples, who can give respectability to their papers by referring back to the earlier published literature. In other words, the referee system, for all its apparent rigour and scrupulosity, by no means closes the official publication media of science against all irrationality. The assertion made by many cranks, that modern science is completely ecclesiastical, orthodox, and impervious to revolutionary criticism, is by no means fully justified. Notice, however, that serious attention is seldom given to pathological science unless it comes from some apparently reputable source. Blondlot was a member of the Académie des Sciences; Davis was a professor at Columbia; Barnes and Allison were evidently working in academic laboratories; Rhine, as everyone knows, is a professor at Duke University. The fact that they had each, at some earlier period, done normal scientific work, and acquired their Ph.D. licence to practise, gave them a far more sympathetic hearing than is normally accorded to the common-or-garden crank without any formal scientific qualifications. As I have already remarked in the case of 'Piltdown Man', membership of the scientific community commands moral respect and intellectual acquiescence from other scientists in proportion to one's status and reputation. The supposed humility and equality of all in the sight of God and Nature is very far from the reality of the scientific life.

I would go further, and say that a significant fraction of the ordinary scientific literature in any field is concerned with essentially irrational theories put forward by a few well-established scholars who have lost touch with reality. Not only do their papers get published, however illogical or incomprehensible; younger, less experienced research workers take them seriously, refer to them, and add their own contributions. If scientific progress is conceived of as mechanical and automatic, then this phenomenon is indeed dysfunctional and pathological.

In more human terms, it is a pathetic symptom of the basic toughness and uncertainty of the scientific life, even for those who are outwardly successful. The admirable paper by Kubie,[1] to which I have already referred, is an attempt to analyse the deep neurotic forces that drive many gifted people into scientific research, and yet give them no happiness therein. Without accepting every one of his psychoanalytic interpretations, we can all provide similar examples from our own personal acquaintance, and can agree with his basic conclusions. The idea that the scientific community is a refined form of the island of Circe, where *all* are blissfully bewitched, is very far from the truth. This is not the sort of fact that comes automatically from a purely sociological survey, because the tortured souls are a small minority, and take good care to hide their anguish – but I think it should be said aloud, all the same.

Or is this sprinkling of neurasthenia, neurosis, paranoia, megalomania and psychosis merely the normal human condition? It could be, by a strange paradox, that the scientific community is much more tolerant of irrational behaviour than the ordinary, busy, practical world. When a bank manager goes off his head, introduces himself as Nubar Gulbenkian, and begins handing out five-pound notes to passers by, then we know that something must be done about it and we quickly put him into hospital. But when the powerful intellect of a distinguished scholar – the man I have in mind is now long dead, but I will not name him – clothes his fantasies in sober garments of sophisticated logic, jazzed up with all sorts of brilliant technical devices, it takes a bold and clever critic to assert that it is all nonsense. Anyway – he was only an old don, gone a bit gaga, so what harm could it do to humour the old boy! Beyond a certain point it is easier, and kinder, to let the fantasies get into print, and then to ignore them, rather than go to the trouble of openly combating them.

Open Combat! Open Controversy! Surprising as it may seem to the readers of *The Double Helix,*[5] this is a state that normal science deliberately shuns. The conflict of ideas, results, claims and theories is basic to scientific progress, but there are powerful norms forbidding intense dispute, personal abuse, alienation and fragmentation into mutually irreconcilable 'schools'. If knowledge is to be 'consensible' then division into groups that do not share common aims, and do not speak or listen to one another, is the negation of science, and must be avoided at all costs. Controversy of this violence, which does occasionally occur, must be regarded as deeply pathological.

Indeed, reading Hagstrom's excellent analysis of the conduct of scientific disputes,[6] one wonders along with him how they ever occur at all! Here is another point at which a description in the language of pure social anthropology – function, role, norms, etc. – does not tell the whole story. Hagstrom concludes his discussion with the bemused comment: 'Scientists appear to have strong personal needs for unforced agreement from others, and may avoid any evidence that such agreement cannot be obtained.'

But that is the whole point. It is easy enough to have a *new* idea. It is easy enough to reject other people's new ideas. The difficulty of science is to have such *good* new ideas that other people will agree with them.

The psychological balance of scientific innovators is exceedingly delicate. One must have sufficient confidence in one's own notions to carry conviction in argument. Yet one must not become so deeply committed that one cannot escape from them if they prove untenable. *The Double Helix*[5] brings out the passion and anguish with which scientific research is really pursued. Kubie remarks that 'the structure of science adds layer on layer, each burdened by more subtle and complex unconscious emotional investments, demanding of the scientist an ever greater clarity about the role of his own unconscious processes in his conscious theories and experiments, and each requiring an ever more rigorous correction for the influence of unconscious preconceptions'. I would add that the 'norm' demanding impersonality of expression – for example, the use of a neutral tone and passive voice in scientific writing – is more than a rule for protecting the scientific community from violent disputes; it is a device for warning us of the personal psychological danger of allowing our private passions to take control of us in public.

Intellectual property is, after all, so intangible and evanescent. An original theory or discovery that is acclaimed today may turn out tomorrow to be an error or a blind alley. The question of ownership is much more uncertain than with patents or copyrights; absolute priority of discovery is meaningless with a slowly evolving complex of ideas; yet a single sentence might be enough to convey the essence of a great revolution of thought. Even the most successful scholar is aware of the danger of losing all the intellectual capital that he has so laboriously amassed. An attack upon his point of view threatens his whole reputation and self-esteem: he sees his name in the history books being erased from the column of heroes, and put in a footnote amongst those tiresome fools who have wasted the time of their contemporaries with their errors and irrelevancies.

It is sometimes supposed that the problems of the scientific life – if not of all living – arise only in youth, under the competitive pressure to prove oneself worthy and to get established professionally. The bitterness of this struggle in nineteenth-century Germany is expressed to the full in Weber's marvellous essay on *Science as a Vocation.*[7] But Weber asks deeper questions. 'Every scientific "fulfilment" raises new "questions"; it *asks* to be "surpassed" and outdated. Whoever wishes to serve science has to resign himself to this fact . . . Why does one engage in doing something that in reality never comes, and never can come to an end?' As he then demonstrates, this conundrum has no satisfactory answer, in logical, mystical, religious or aesthetic terms. But the failure to ask it at all, and thus come to terms with reality, may be a significant cause of bitterness in scientific controversies – a pathological symptom from which not even the elderly and eminent are quite immune. This is a disease that could, perhaps, to some extent be mitigated, if not cured, by a dose of serious moral philosophy – a realm of discourse which is, of course, entirely absent from our whole scientific curriculum.

It is clear from these remarks that we cannot understand the workings of the scientific community, whether in health or in sickness, until we know much more about the *motives* leading to a scientific career. This question is shrouded in the deepest obscurity; I should hesitate even to quote the titles of the books and essays that have been written on the subject, for I think they have demonstrated nothing for sure. Kubie, for example, talks of the need to gratify unconscious neurotic desires; by contrast Hudson[8] seems to discuss inborn traits of 'convergence' and 'divergence' of mind. My own observations would point to sheer social inertia – the academically able young man who finds himself at university reading science nowadays follows the line of least resistance by applying for a research grant and taking a Ph.D., just as he would, a century ago, have gone into the Church.

I would refer, also, to that clever little book *The Struggles of Albert*

Woods[9] which demonstrates the power of sheer vanity in an intelligent, energetic, if common-place, personality. The vulgar, practical answer to Weber's deep question – an answer that he firmly rejects – is that, like any good game, it gives one a chance of showing off, and winning a round of applause, before the whistle is blown.

If we did have a sound understanding of the motivation of science, then we could begin to think about *incentives.* Are the rewards of scientific achievement properly matched to the social psychology of research? This is an important question, for the material rewards nowadays can be very great. Although we cannot put our finger on particularly damaging episodes, many of us are worried about the 'Nobel Prize complex' that seems to govern the activities of so many of the best modern scientists. We are undoubtedly witnessing an intensification of a 'star' system, where tremendous fame, power, intellectual authority, and even wealth, are going to a few fortunate individuals. It is difficult to judge the effect of this on the whole enterprise. Does the intensified competition really produce more intellectual goods for the manpower employed? Would Francis Crick and Jim Watson have devoted less effort to the search for the structure of DNA if there had not been the possibility of a Nobel Prize at the end of the trail? Or do such prizes lead merely to the adventurism, grossly speculative theories and doubtful experimental claims that we sometimes observe? Are not scientific discoveries of that quality their own reward, without this extra material bonus?

The pathological phenomena to which I would draw attention are the contrived, and yet almost fortuitous, titles of eminence within the very top ranks of the scientific community. A few hundred Nobel Laureates are not enough to go round amongst the several thousand living scientists who are, one would say, in that class of ability. The same applies to membership of the National Academies – even to our own dear old Royal Society – which have not expanded in numbers to match the growth of the scientific community. Election becomes more difficult – and the disappointment of reasonable expectations is that much more bitter. I fully appreciate the difficulty of any public discussion of this sort of thing, but I think we must admit that it is unhealthy when we can point to a number of first-class scientists, still in the prime of their powers, whom everyone agrees to be fully of the supposed standard, and who are yet unlikely now to be elected. The function of the Royal Society, and of similar Academies, is to be a meeting place and organ of opinion for *all* the most responsible and senior scientists in the country, and not just a prestigious labelling machine for those who have been fortunate enough to get their noses slightly ahead at the right moment in the race.

What we see, in these public prizes and distinctions, is yet another attempt to avoid the consequences of Weber's question. Scientific achievement is an intangible, wasting asset; let us replace it with definite, permanent symbols of social esteem. The German professoriate took their rewards in State Honours and Privy Counsellorships; in the Soviet Union there are extra privileges and special comforts for Academicians; in America, the symbol of success is money – $50 000 or so a year to solace one's declining years as a 'Meritorious Professor'. This disease, all too human and forgivable, may be relatively mild, yet it is endemic in the scholarly community and should not be regarded as quite harmless.

Am I insisting, puritanically, that the scientist should lead a lean and austere life? What I mean is not that we should all live like hermits, but that great affluence should not come too easily. It is difficult to point to specific and scandalous failures – yet one must be aware of a certain flabbiness in the intellectual world due to the enormous subsidies it has had to digest. In recent years, it has become just a bit too easy to get a Ph.D. of sorts, and an academic job of sorts, and a grant for research of sorts – to produce yet more Ph.D.s of the same sort. Scientific research is not simply a professional job, to be entered into in the spirit of plumbing or accountancy. Nor is it a mechanical activity, like programming a computer, that can be learnt from a manual. It demands a sense of vocation and dedication; it is an art form, transmitted from master to apprentice by oral tradition and imitation over a period of years. To expand the scientific labour force more quickly than it can be adequately trained must surely lower the quality of the product. The current shake-out in the Ph.D. job market is not, to my mind, a symptom of some serious disease, but rather an occasion for a nature cure by gentle dieting! Without wishing to return to the bitterly competitive situation of the 40-year-old *Privatdozent* of Weber's time, I believe that a few years of struggle, in which quite a proportion of would-be research workers could discover their real talents as teachers, engineers, managers or salesmen, would purge the scientific community of many useless members who have not the faintest notion of what it is all about.

The fact is that the academic community, despite its tendency to look inward and not care much about what the world is doing, is quite easily bought and sold. Dons may not be very ready to undertake the dirty work of Industry and Defence, but they are quite willing to be subsidized to any extent, in the name of basic research, etc., by those whom they despise. Up to now I have tried to discuss symptoms of dysfunction within the scientific world itself, to exemplify the tensions between social norms and psychological drives. We should now consider briefly the pressures that are exerted on individuals and institutions by the demands of society at large. In the simplified ethics of the Socially Responsible Scientist, these should be paramount, but the true situation is not all that clear.

This is really an enormous subject. Let me give two examples, from very different contexts. In the journal *Science* there appeared last year an article 'concerning the development of amorphous semiconductor switching devices'.[10] For connoisseurs of scientific controversy, this is a choice item, with many interesting aspects. One episode may be summarized as follows. A reputable basic physics journal refused publication of a paper claiming important discoveries in a field relevant to a new technical device. This decision was later reversed, chiefly as a result of letters from two leading academic scientists arguing that this work was of great interest to the international physics community. The day before the paper appeared, a press conference and other publicity gave rise to many favourable comments which had a strong influence on the stock market valuation of the company proposing to produce this device. Both the professors concerned were consultants to this company, and one of them owns some stock.

Nothing out of the ordinary, you may say; the sort of thing that happens every day in the world of business; why shouldn't the scientists take a small share of the immense wealth they create? It is likely, indeed, that they saw this as merely part of the continual game of wits between author and referee, or even suspected that pressure was coming from a rival company and wished to see justice done to their own side.

Yes, it is only a trivial episode; but I see it as a symptom of a serious disease. The point is that a college professor, in his *scientific* activities, is assumed to be entirely disinterested. When he acts as a referee of papers, as editor of a journal, as the author of a book or review, as organizer of a conference, or as supervisor of graduate students, it must be taken for granted that he offers opinions, gives advice and makes decisions, entirely on the scientific merits of the case. As I have already remarked, good faith and mutual trust are absolute preconditions for the scientific enterprise. The merest suspicion of venality, or personal interest, destroys his whole claim to expertise. The pathological feature in this case is that the professors do not seem to have disclosed their personal interest when writing to the editor of the journal. The academic scientist who is drawn into the industrial, commercial or political world certainly does not lose his reputation as a scholar by becoming a partisan of a particular corporation or organization, but his expert scientific opinion becomes worthless if he does not make it very clear, on each occasion, for whom he is really speaking. A Hippocratic Oath for scientists should put this principle amongst the very first. If it were not well understood, then the whole scientific community would be wide open to the terrible plagues of bribery and corruption.

One can go further, to argue that a whole discipline is in a pathological condition if all its leading exponents are thus tied to commercial or government interests outside the professional research community. The situation that is said to have arisen after the Santa Barbara oil leak is instructive: independent experts could not be found to assess the degree of culpability of the owners of the well because all the leading petroleum geologists were already being retained as consultants by the big oil companies. When I mentioned this to an officer of the National Science Foundation he said 'Oh yes – we would send any proposal for research in that field over to the oil companies; they would support it as a matter of course!' Much of the apparent opposition to making scientists perfectly 'responsible' and 'relevant' in all their research is based upon the paradox that science must protect itself from many of the pressures of society if it is to exert its full power as a responsible force in social affairs. Perhaps this is only too obvious; yet it is a conjecture of sufficient importance to merit support by proper sociological analysis.

My other example, of a pathological – indeed totally destructive – effect of social pressure on organized science comes from the other political pole – from the Soviet Union. I refer, of course, to the tragic phenomenon of Lysenkoism. We must all be grateful to Dr Zhores Medvedev for writing, and allowing to be translated and published abroad, his vivid and authentic history[11] of this terrifying aberration in a country supposedly dedicated to rational scientific materialism. The terrifying feature was the apparent normality of the scientific organization within which this false doctrine existed. Visiting the Soviet Union ten years ago, one would have found great research institutes, directed by Academicians and professors, manned by hundreds of graduates in botany, zoology and agriculture, fully equipped with microscopes and trial grounds, incubators and scalpels, publishing hundreds of scientific papers, advising farmers, holding conferences, and behaving just like other scientists – with a recognizable, testable, nonsensical lie at the heart of it. The tragedy lies not with those few who were forced, by irresistible threats of violence, to keep silent or to say the opposite of their thoughts; it is with the many who seem to have accepted the doctrine as it was taught to them and failed to subject it to their own tests of rationality. We cannot blame a half-educated crank for believing his own theories and trying to get them accepted; we must ask what was wrong with a whole scientific community that it allowed itself to be captured by such crazy notions.

I cannot do better than quote Medvedev's own answer to this question:

> False doctrines, being an extreme product of the normal background of science, and having been created by extremist fanatical representatives of the world of science, can achieve a monopolistic position only in state systems that are extremist in nature, as a particular manifestation of many other deviations from the reasonable norms of organised human society.

He then goes on to discuss more specific causes – the tendency to classify science as 'bourgeois' or 'proletarian'; difficulties and mistakes in agricultural policy; press censorship; the isolation of Soviet scientists from world science; and the rigid centralization of scientific education, publication and administration.

It is easy enough for us to say that it could not happen here. We should, indeed, be singularly unfortunate to have full-blooded Stalinism wished upon us. I hope that at least a few of our own scholars would then behave as heroically as some of the Russian geneticists in those terrible years. But I think we can learn, from this as from the other pathologies I have discussed, some positive facts about the scientific community and the way it really works. In our own democratic and liberal social environment, we take for granted such elementary features as freedom of speech, freedom to travel, permanent tenure of academic office, an independent diversity of universities and other institutions, which are the real guarantees of healthy, productive science.[12]

In his Presidential Address to Section N of the British Association, at Leeds in 1967, Edward Shils[13] spoke of the Profession of Science, drawing attention to the self-sustaining and self-governing character of this community which sets it apart from other professions such as medicine or law. He also spoke of the problems that arise, especially in the choice of priorities, when demands are made upon science by industry, defence, and other social organs. I agree fully with his argument that the institutionalization of science was essential for its development to its full powers, but I am concerned lest an inappropriate social structure should then be imposed upon the scientific community by an insensitive governmental machine. It is just not good enough to set up a bureaucratic hierarchy, hire a lot of chaps with Ph.D.s, put them down in a beautiful new building full of shining equipment, and tell them to get weaving. Given an existing tradition of scientific excellence and a clearly defined objective such as winning a war or going to the Moon, this sort of military or factory production procedure can work for a while – but if it is to continue to prosper it must be sustained by the inner spirit of Natural Philosophy and by the ancient courtesies of the Invisible College.

The plain fact is that the scientific community has failed to produce, and inculcate in its members, any conscious institutional ethic. Joseph Haberer,[14] in his perspicuous study of physics in Germany before and under the Nazis, and in the United States during the Oppenheimer outrage, gives ample evidence for his judgement that 'scientists have almost always been pliant partners [of the State], willing under almost any conditions to accommodate to a given political order'. But the policy, as he calls it, of 'prudential acquiescence', besides being ignoble is disastrous. To quote Haberer again:

> Prudence and a capacity for compromise are frequently a necessary, even a desirable, characteristic among leaders. But when the integrity (perhaps even the very existence) of an enterprise or professional community is seriously endangered, a leadership that ignores the fundamental commitments and obligations inherent in the life of a community becomes an added danger, and contributes to the eventual disaster that is likely to overtake it. In such a time of crisis, to relegate the defence of fundamental principles to a secondary position and to champion instead a policy founded in expediency and compliance, is to acquiesce in the threat to the existence of the community.

I am *not* saying that the whole body of science is now unhealthy and corrupt, that we are in a state of deep crisis, and must take desperate defensive action against the terrible forces of a new Dark Age. Indeed, as Medvedev's book shows, the Soviet geneticists were by no means acquiescent; the scientific community, despite its weakness and incoherence, has an inherent self-healing power, through the close contact of its members with hard material reality in the laboratory. But I think there is an important lesson to be learnt in this example of a great institution that was driven into a completely psychotic state, despite all the money and attention that was lavished on it by the Soviet Government.

At the beginning of this Address, I was somewhat apologetic for the unprofessional subject matter and amateur manner of treatment. Now, at the end, I do not ask that it be considered a work of science or scholarship.

The questions are badly posed, the evidence is qualitative and fragmentary, and the conclusions arrived at are trivial or trite. Yet I still think that this is a subject that ought to be pursued seriously; not to debunk science, but to understand it, and to love it, 'warts and all'.

References

1. Kubie, L. S. (1962) *Some Unsolved Problems of the Scientific Career.* Reprinted in *The Sociology of Science,* edited by B. Barber and W. Hirsch, pp. 201–29. New York: Free Press of Glencoe.
2. Ziman, J. M. (1968) *Public Knowledge.* Cambridge University Press.
3. Snow, C. P. (1958) *The Search.* London: Macmillan.
4. Langmuir, I. (1968) *Pathological Science* – the transcript of a talk given in 1953, published as a report in the Technical Information Series of the General Electric Research and Development Center, Schenectady, New York.
5. Watson, J. D. (1965) *The Double Helix.* London: Weidenfeld & Nicolson.
6. Hagstrom, W. O. (1965) *The Scientific Community,* p.278. New York: Basic Books.
7. Weber, M. (1965) *Science as a Vocation.* Reprinted in *From Max Weber: Essays in Sociology,* edited by H. H. Gerth and C. Wright Mills. New York: Oxford University Press.
8. Hudson, L. (1966) *Contrary Imaginations.* London: Methuen.
9. Cooper, W. (1952) *The Struggles of Albert Woods.* London: Jonathan Cape.
10. 'Ovshinksy: Promoter or Persecuted Genius', *Science* 165, 673 (1969).
11. Medvedev, Z. A. (1969) *The Rise and Fall of T. D. Lysenko.* Translated by I. Michael Lerner. Columbia University Press.
12. *See, especially,* Polanyi, M. (1968) *Personal Knowledge.* London: Routledge & Kegan Paul.
13. Shils, E. (1968) The Profession of Science. *Advt. Sci.* 25, 469.
14. Haberer, J. (1969) *Politics and the Community of Science.* New York: Van Nostrand.

20

Why be a Scientist?*

The social context of research cannot be ignored.

Science as a vocation

The aspiring young scientist dreams of 'pushing back the frontiers of knowledge'. There is a mysterious desire to 'explore Nature' or to 'satisfy human curiosity'. Pressed to give a more precise motive, the aspirant may murmur something about 'making two blades of grass grow where one grew before'. For some there may be an unacknowledgeable ambition to win fame and fortune with a Nobel Prize. For others it may be just a

Figure 1

* Address to Section X of the British Association, September 1979.

wonderful opportunity to practise some absorbing craft, like mathematical manipulation, electronic design, or messing about with bubbly coloured liquids and glassware. Some take to science because there are jolly good at it and have no confidence that they could succeed anywhere but in academia. For others, the vocational awakening can come almost too late, as they stand on the very verge of the plunge into real life.

Does it help very much to discover some order of priority amongst these various motives? They are all reputable enough in their various ways, and few scientists could tell you, after the most conscientious heart-searchings, which were uppermost in their own original choice of a career. Indeed, the very idea of a conscious choice may be misleading. The science education machine (Figure 2) carries them up from level to level of 'valid' science. At each stage, at an appropriate age, there is a barrier to be passed. Those who succeed easily are given every encouragement to go on; those who fail, or falter, move off into real life, or to technical work, or to technology, so that only a chosen few enter that idealized realm of research. Notice the part played by the teacher in this screening process. Having themselves reached a somewhat higher level than the courses they teach, they communicate this aspiration to the best of their pupils, largely ignoring the major proportion who will, in fact, move out of the system at an earlier stage.

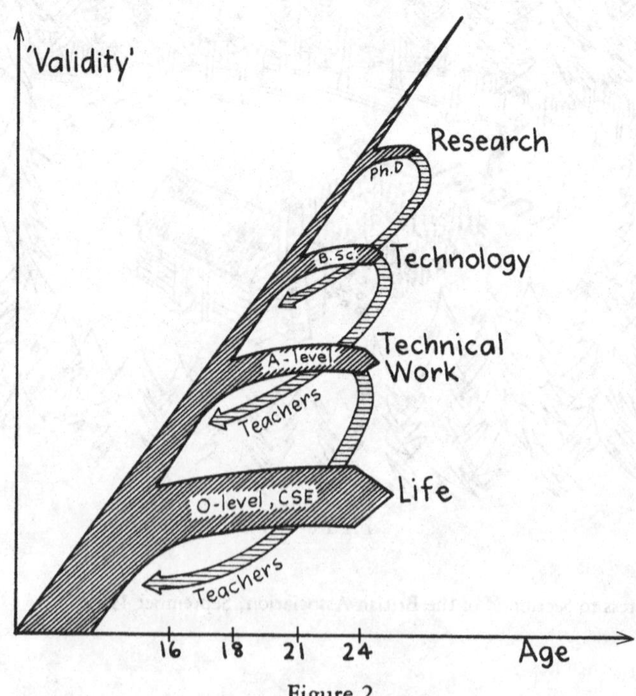

Figure 2

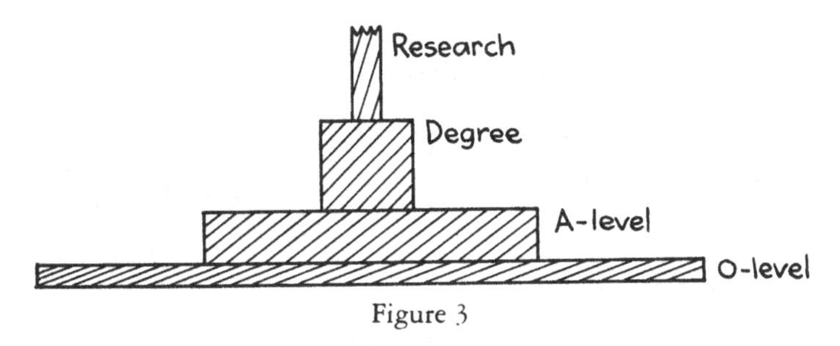

Figure 3

For each pupil the curriculum of a valid science looks like a pyramid (Figure 3). The base at O-level is very broad and very thin. At A-level it thickens and narrows. An Honours degree is no more than one discipline wide. The student who gets through a research degree is a true specialist, having learnt a hundred times more about a thousand times less than any ordinary person would want to know. This is what is needed to get to the research frontier. The continued progress of valid science depends upon the division of the labour of research into a thousand problems, sub-sub-fields and research projects. Not only have philosophy, humanities, religious studies, politics, economics and most time-consuming recreations been thrown overboard; there will be only a vague memory of some traces of some of the other sciences that might possibly have been helpful in research but which have had to be dropped in earlier years.

The little world of the science don

The apprentice research worker, with his new-won Ph.D., lives in a very limited world. It is not exactly that it is one-dimensional. The scientific life cannot be lived entirely according to the stereotype of an individual personal role, or as 'one undergoing education', or as the practice of a technique, or as the plaything of the political forces, or as the manifestation of an intellectual process, or even as just a member of some human group. The richness of life is that each of these is no more than an aspect, or a dimension, of our total being (Figure 4). But in that multi-dimensional space the world of the research scientist is highly circumscribed. Along the educational dimension there is only the Graduate School. The only technique is that of Basic Research. The only interesting political question is whether the grant will come from the Research Council. The only significant cognitive elements are scientific Concepts, and the only group worth belonging to is the Invisible College of the subject. Outside of this compact and bounded territory – it might be an austere life on the

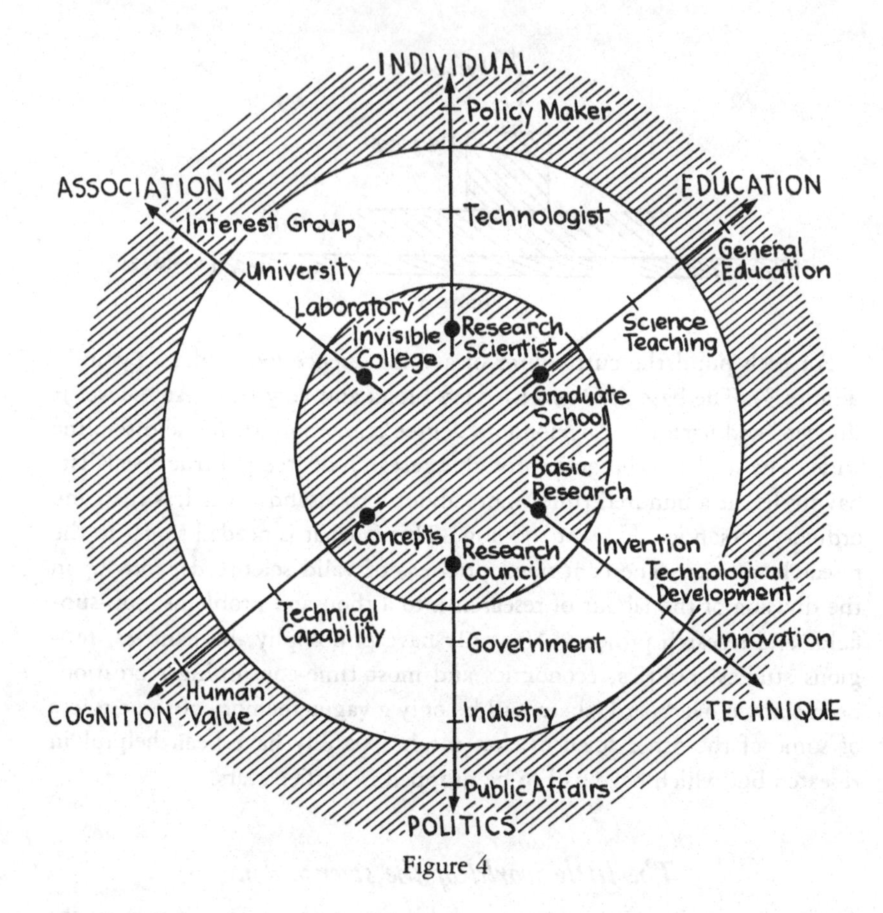

Figure 4

Arctic ice, in the domains surrounding the 'North Pole' of a hypothetical hypersphere – there are only distractions and uglifications unworthy of concern.

The three dimensions of academic science

To the lay person the world of science has a simple enough structure (Figure 5). Scientists spend happy busy lives doing research. The results of their observations, experiments and theories are then communicated, in various ways, to a grateful world. The knowledge thus obtained is written up in great fat books and stored in public archives such as university libraries. Perhaps there was a time, long ago, when this was a good enough description of what it meant to be a scientist, although, of course, this entirely ignores the social, technical and metaphysical ocean in which each individual 'natural philosopher' was completely immersed.

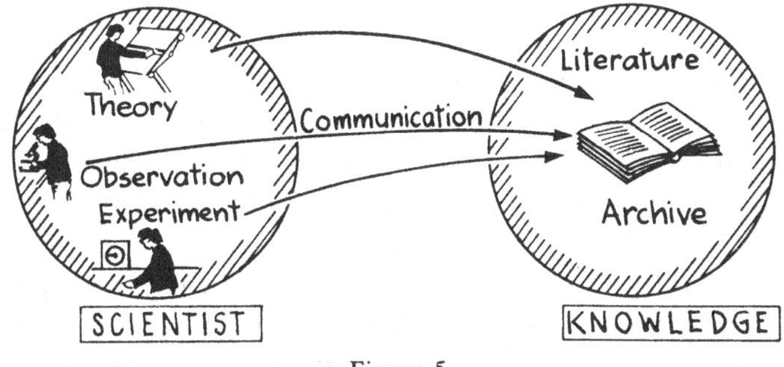

Figure 5

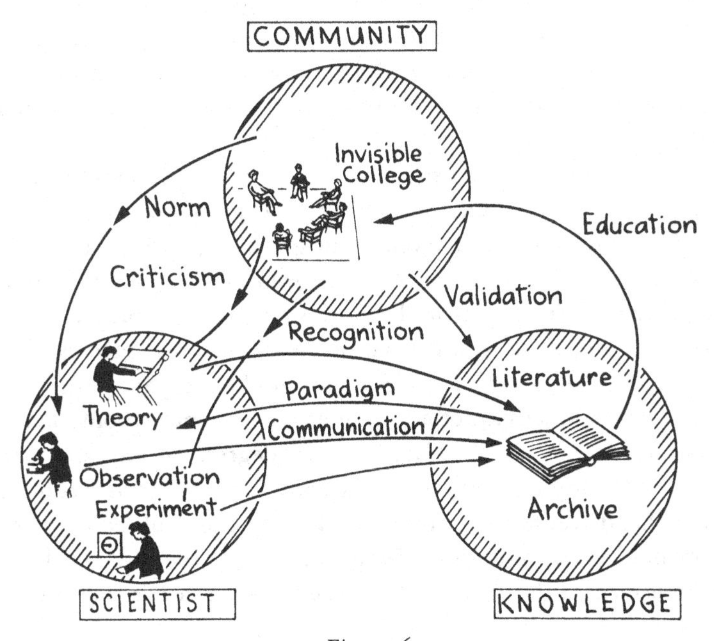

Figure 6

But from about the beginning of the eighteenth century one cannot ignore the communal aspect of European science. The academic style of science (Figure 6) involves each individual scientist in competitive and cooperative relationship with other members of the scientific community. This is the human group that the graduate student enters on the way to a Ph.D. Learning to be a scientist is not just a matter of building apparatus, doing experiments, reading up the background of the subject, and inventing theoretical interpretations for what has been observed; it is also a matter of learning to accept the criticism of one's research by other scientists, and picking up all sorts of protocols and norms of behaviour as a member of the scientific community. One has to learn, for example, to suppress the personal pronoun in formal communication, such as scientific papers or lectures, and yet somehow indicate that one is a lively imaginative person full of bright ideas. There are moments of anger and exasperation with the obtuseness and stupidity of other scientists – but controversy must be conducted with the utmost courtesy and diplomatic impersonality. There should be moments in which one is deeply chagrined at having had an error corrected in public – but that teaches the lesson to be scrupulously careful and self-critical in one's own work. The popular conception of research is that it is a struggle to master mysterious Nature; it turns out, in many instances, to be an equally painful struggle to master one's own wilder nature and yet not to be mastered altogether by the greater authorities in one's subject, or by much too clever contemporaries.

Nevertheless, there are also emotional rewards in belonging to the Invisible College of one's specialized subject. It is gratifying to get a paper accepted for publication. It is even more gratifying to have it cited by others. Then there are opportunities at conferences to expound the results of research and to discover that there are quite a lot of pleasant people who are really interested in just what one is doing. This process of 'recognition' can raise one's ego to giddier and giddier heights – the invited lectureship or chairmanship of a session at a conference, promotion to a Chair, election to a great academy, honorary degrees, awards and prizes. It would be misleading to exaggerate the extent to which this sort of fame is the spur that drives many scientists on. But it is, undoubtedly, essential for the mental health of every research worker that there should be tangible evidence that other people really care for what he, or she, is doing, and think it good.

Imagination versus criticism

The relationship between the individual research worker and the archive of knowledge is also somewhat more complicated than many people realize. There is quite a long and rigorous process of refereeing, critical re-evaluation, further experimental testing, etc., by the community as a whole before any particular scientific result is regarded as fully established and valid. In fact, it is by this untidy, anarchical and sometimes gravely imperfect process that scientific knowledge slowly gets purified of error and fallacy. Research scientists have practically no use for formal logic.

They work within the epistemological traditions of their subject and persuade each other as best they can that they have arrived at the truth.

On the other hand, the knowledge in the archive is not just a passive reservoir to be tapped for technical application. It forms the intellectual atmosphere in which new research must proceed. Concepts that were revolutionary and speculative become, in due course, the paradigm of the subject for the next generation. The Young Turks that made the revolution may themselves need to be overthrown, as further progress is made. The real intellectual attraction of science is the tremendous tension between conservative and radical forces: the conservatives, to maintain and strengthen what is sound and reliable and well established, which is always in danger of being forgotten, or overlain with fallacious novelties; the radical trend, to discover the misconceptions in what is taken for granted or to break into quite new realms of thought with new experiments or new theories. In many walks of life these forces are in destructive opposition. But in science the dialectic is truly creative. The conservative-minded scientist knows that much of what he would like to preserve is inadequately tested and that a well proven innovation will deserve its place; the radical knows that he must put his feet on firm foundations of fact or interpretation, even if he is going to pull down and rebuild part of the edifice. One of the most important stages in the maturing of a scientist is the realization that the path to truth must pass between these two extremes – and perhaps the acquisition of sufficient self-knowledge to know his, or her, natural tendencies as between the critical and imaginative cast of mind.

Acceptance into, and socialization by, the scientific community does not come about just by the award of the Ph.D. It is quite a slow process, taking perhaps five or ten years, from the apprentice research student to the fully confident and intellectually independent research scientist. There is an inner psychological development of mastering skills and techniques, internalizing public norms of behaviour and intellectual standards, discerning attainable research goals and setting for oneself a research programme that may turn into a life's work. The most desirable attribute of a young scientist is to be 'self-winding' in the planning, the execution and the communication of research. And, of course, it takes time to build up a reputation with a good record of 'contributions' to science in the form of published papers. There is also a movement outward from the very cosy and enclosed world of a research group – usually under the fatherly eye of a benevolent but very powerful patron, in the form of a professor or research group leader – into the national and international circles of world science. There may be a real need to break away from local patronage, local interpretations of the communal norms, and local paradigms of research method and goals, before this maturation process can be completed. Academic mobility that brings new people into new institutions may be as important for the individual curriculum vitae as for the institution to which he or she brings new blood.

Industrialized science

Academic science, as thus described, can be a delightful life for those who measure up to its demands. In fact, even if you don't quite reach the final stage of scientific maturity, once you have tenure you can still make quite a go of it as an academic, or as a full-time research worker in an industrial or government laboratory. Unfortunately, this style of science is now somewhat old-fashioned and is giving way to what can only be called 'industrialized' science (Figure 7). This has come about quite simply, in the last few decades, by the overwhelming demands for very elaborate and expensive scientific apparatus. There is no need to describe at length the immense scale on which some of the facilities and instruments of 'Big Science' are now constructed. Particle accelerators and telescopes, space satellites and oceanographic vessels, cost hundreds of millions of pounds at a time. Similar sums aggregate in the many separate pieces of the equipment, such as spectrometers, electron microscopes and animal houses, that must be brought together to support research in any modern scientific discipline. This is the way that modern science makes progress, and there is really no return to more primitive styles.

It is not so much a question of getting the money for these facilities. From the point of view of the individual scientists, it is the effect on the research life itself. It is essential, for example, that research should be carried out in organized teams, where as many as 20 or 30 fully qualified scientists (that is, with Ph.D.s) must work together on a single project, subordinating to its common purpose both their own originality and their personal prospects of recognition and reward. Such work is, of course, fully demanding of all the qualities of intelligence and technical skill that we expect of the independent research worker – but its norms and goals are much more akin to those of the engineer or other highly skilled technical worker in a corporate organization than they are to the traditional motives and social relations of the scientific community. That is to say, the balance of the scientific life shifts more towards technical work, away

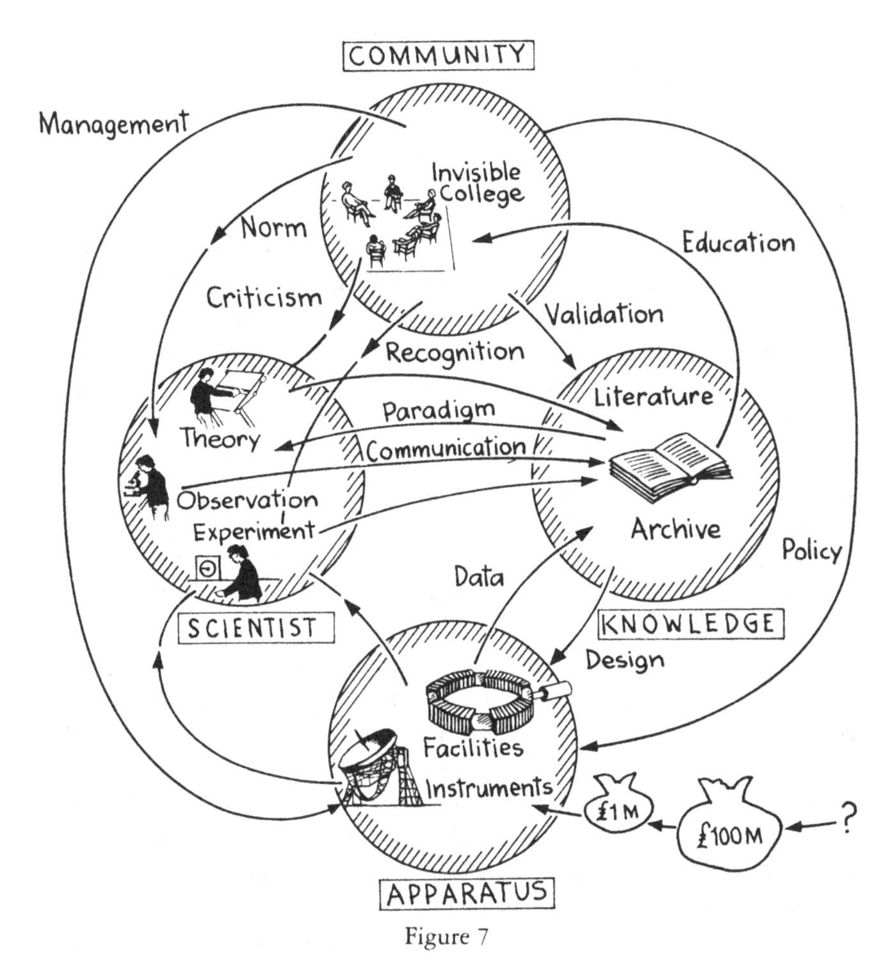

Figure 7

from natural philosophy. The archetypal scientist was a self-winding individualist; now the stereotype is an Organization Person performing a prescribed task to the satisfaction of colleagues and bosses.

The work itself changes in detail. Much more effort must be brought into the design and construction of ever more elaborate instruments, whose performance is largely preconceived, and much less into imagining cunning or crazy experiments by which Nature might be caught napping. The data are no longer laboriously noted down in scruffy notebooks and cogitated upon whilst the apparatus is being set up for the next measurement; they come pouring out of the on-line digitalized computerized instruments, already processed into, and out of, machine-readable form so that they could almost be posted off at once for publication in a Letters journal.

Managing R & D

The communal relations of science are also being transformed by this industrialization process. At the level of an institution such as a research laboratory, there has to be much more management, both to do and to be done by. The delicate social norms, within which students learn their craft from their professors and then, in turn, set up as masters, must be supplemented with more rational bureaucratic arrangements, so that it should be quite clear where the responsibility lies for the purchase and use of consumable stores, salary scales of research assistants and technicians, maintenance contracts for major electronic equipment, allocation of rooms, secretaries and computer time, and so on, and so on. The careers of many successful scientists drift from the research mode into the administrative mode of work, simply because that is the way that Big Science must be done.

And at the very top of the system, there is all that budgeting and allocation of funds, and award of research grants, and negotiating with government departments and chairing committees, that come under the general heading of science policy. No doubt it is very gratifying to become a scientific notable, loaded down with all the rewards of recognition by the scientific community. But this may carry with it immense responsibilities, worries, and distractions from the research process itself, as one tries to grapple with all these problems.

The industrialization of academic science has not only transformed it internally into a much more bureaucratic social institution; it has also become much more closely linked with society in general. The boundaries between science and technology have become so open that we now have a general 'R & D' system in which science, technology and education merge into one another almost indistinguishably (Figure 8). A university or polytechnic lecturer may be as much involved in consulting for industry as with teaching science or doing research. Science Policy includes concern about the economics of industrial innovation and the supply of 'QSEs' (Qualified Scientists and Engineers) as well as decisions about the relative funding of academic research in astronomy and molecular biology. Although, in practice, there may not be a great deal of career mobility between industrial, academic and government employment, the scientific life is not so very different in a big university science department from

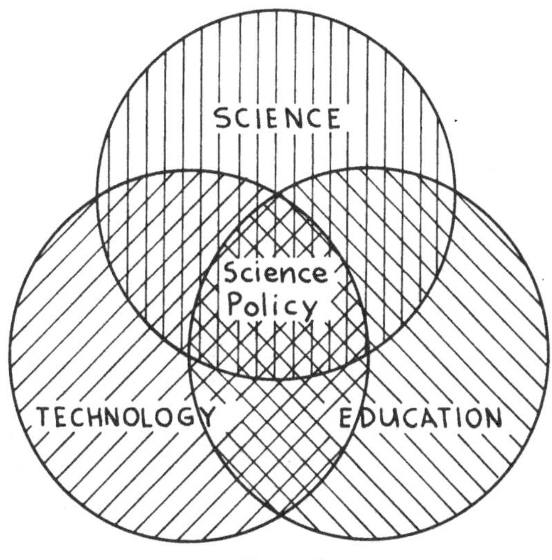

Figure 8

what it is in the more basic research laboratories of a large industrial corporation, or in a government research establishment such as Harwell.

Scientists in their social context

But this system is open to social influences all around the edges (Figure 9). A QSE working in R & D can no longer claim to be just an 'honest seeker after truth'. In three cases out of four he, or she, is serving an industrial corporation with quite definite and worldly goals, such as making profits or weapons of war. The scientist whose research is paid for and used by such a corporation cannot claim 'benefit of clergy'; if one is employed by such an organization one must accept some responsibility for what it does and how its products are used.

By participation in the R & D system a scientist who has acquired intellectual authority by contributions to basic knowledge may be called on to offer expert advice on a variety of issues which are much more in the sphere of politics, economics and social values than in the realm of scientific knowledge. Should that advice be given solely within the framework of what is precisely understood and clearly established scientifically, or is there not some duty to offer more subjective opinion, to give some guidance to those who must take difficult technological decisions in a fog of scientific uncertainty? Which of several competing loyalties should have precedence – loyalty to abstract, absolute 'truth', loyalty to the commu-

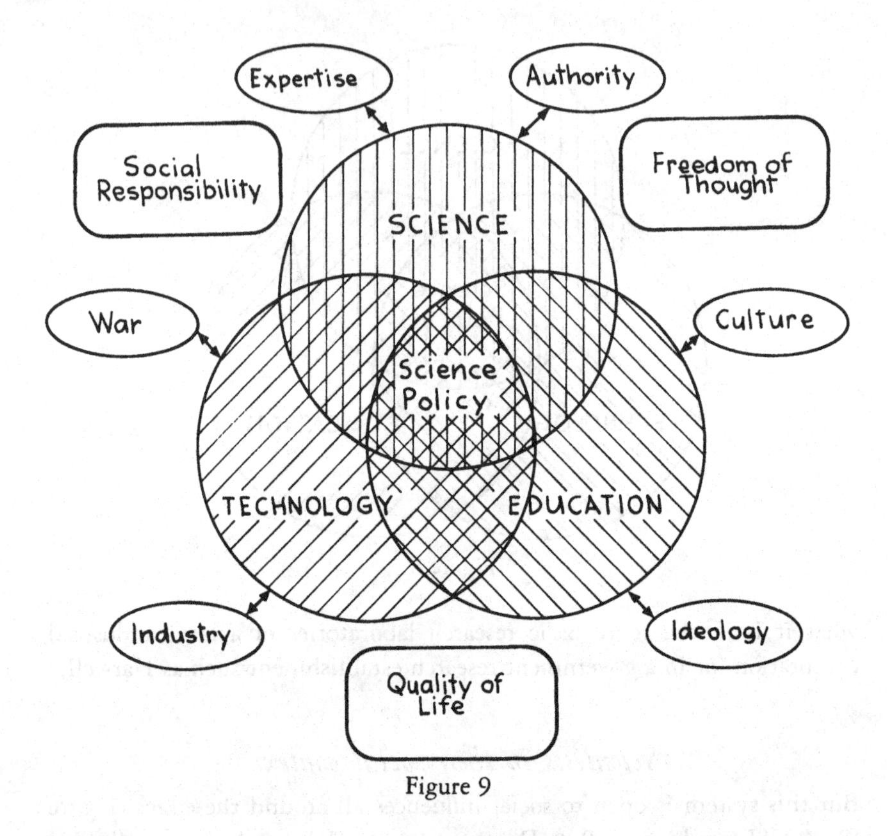

Figure 9

nity of scientists, loyalty to the good of humanity in general, or loyalty to one's country, company, or government department? Such ethical issues may turn out to be much more worrying than any problem encountered in research itself.

Modern culture is permeated with science. But there are great barriers of ignorance and misunderstanding between what is known to individual scientists within their specialities and what is imagined by the public at large. Both scientific and anti-scientific ideologies are active forces in our civilization, and are, of course, heavily influenced by the attitudes and actions of scientists themselves, both in the way they present the scientific world picture and in their responses to particular issues where the quality of life may be at stake. It must be admitted that most scientists choose not to involve themselves in the popularization of science, or in the public debate on controversial scientific or technological questions. Nevertheless, they are uneasily aware that these issues touch upon the central goals and commitments of their own lives, and when they defend themselves by

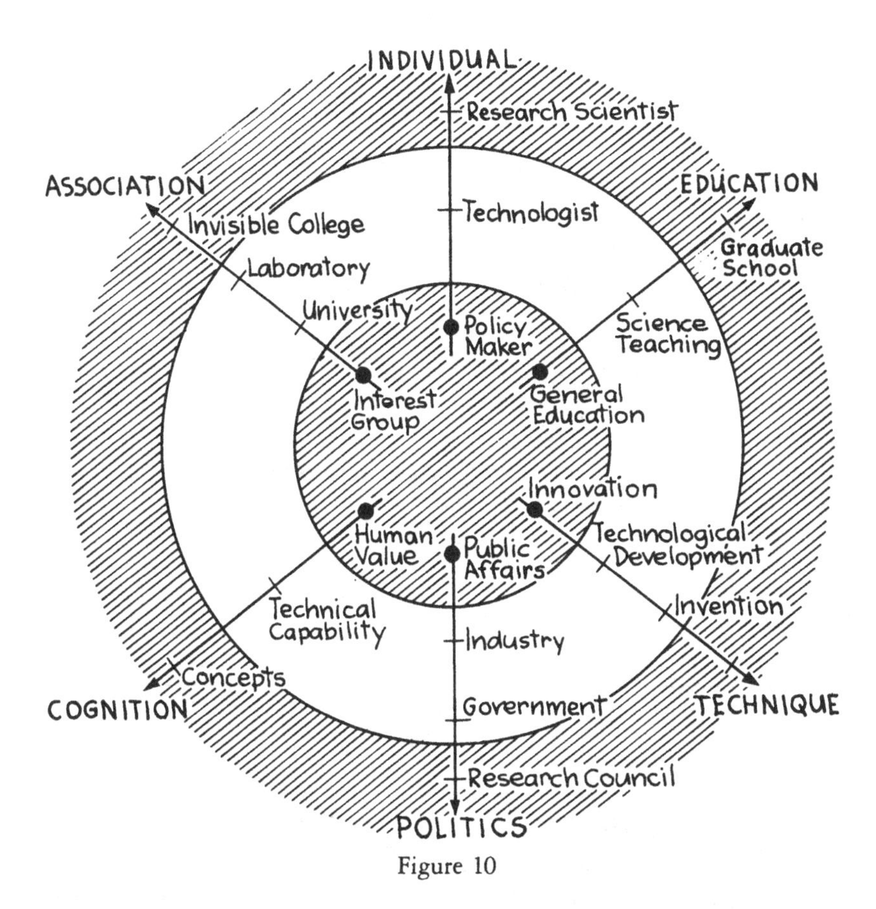

Figure 10

saying 'this is much too complicated and difficult to explain to a non-scientist' they may really be trying to suppress their own doubts about the value and validity of their work.

The experience of actually being a scientist may eventually lead to quite a different view of science itself (Figure 10). It may take one eventually to quite the opposite pole of the hypersphere – to the Antarctic regions, so to speak – where human values draw scientists and other citizens together into interest groups, or involve them with the political, industrial, economic, and military policy makers in the political realm. They may find themselves much less concerned with basic research problems than with the technological innovations they make possible, and their concern in the dimension of education may be as much with general education about science as with what you need to know to get a Ph.D.

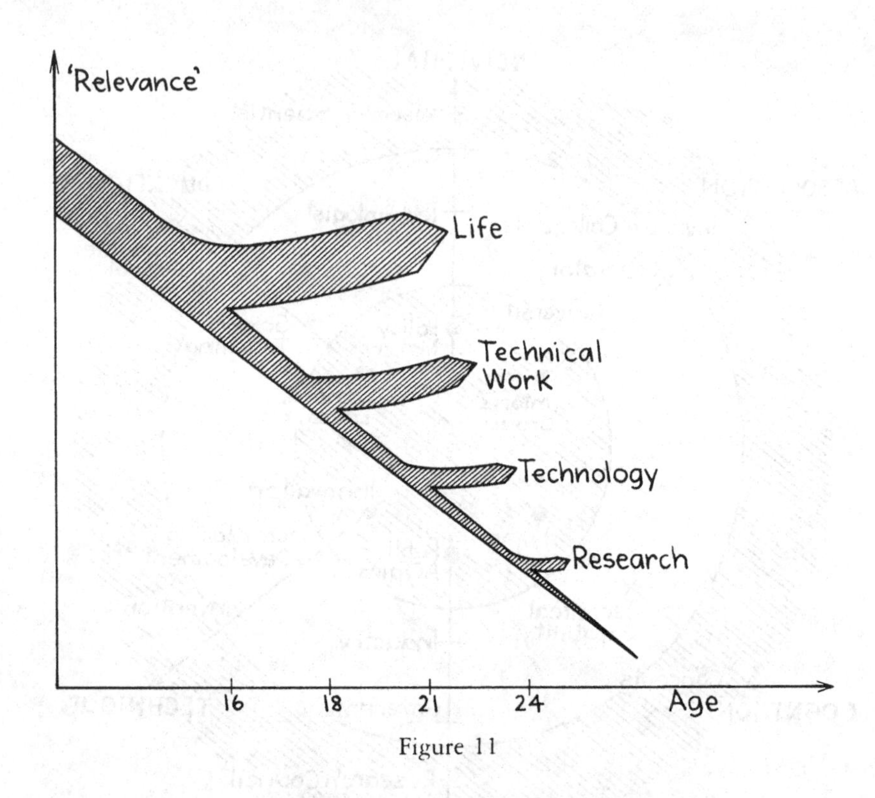

Figure 11

Preparation for a less irrelevant life

Let there be no suggestion that the primary motives of those who become scientists are misconceived. The personal satisfaction we gain from solving a technical puzzle, investigating a scientific problem, or giving free play to curiosity concerning some enigmatic aspect of the great world, are still with us. The rewards of successful research – delight in fine craftsmanship, ecstatic moments of discovery, and acclaim for a genuine achievement by one's peers – are still as gratifying as ever. For all that is said about the competitiveness of modern science, there are plenty of quiet and sheltered spots for those who can't keep up the hot pace of the front runners. As professions go, scientific and technological research is relatively free from petty corruption, malicious gossip, ruthless ambition, and social snobbery. In fact, if it's the sort of life that you can get into by meritocratic achievement and enjoy for the sake of the work itself, it is one of the most delightful and satisfying careers that anybody could have.

Nevertheless, even for those who are well equipped for it by talent, temperament and training, it may not turn out quite as they expected.

Education for valid science, leading onward and upward to the research degree, is a very inadequate preparation for what such a career may eventually entail. Science education is prized for its concentration on objective knowledge, but this concentration is achieved at the expense of increasing irrelevance to all other aspects of human affairs. Take Figure 2 and look at it, so to speak, from the 'South Pole', where most people stand; you might see it then as in Figure 11 – not as a triumphant ascent to a wisdom encompassing all things, but as a process of tunnelling deeper and deeper into narrower and more secret tunnels, away from the freshness and light of real life.

PART FOUR

SCIENCE AND SOCIETY

21

The Impact of Society on Science*

Science must face up to its social responsibilities, although not necessarily by becoming subservient to short-term social demands.

It is not given to us mortals to perceive the full consequences of our actions. Moral responsibility is therefore an issue that cannot be decided by scientific procedures. When an evil is traced back to a cause that we have freely created, we can almost always produce an excuse.

'I had no idea that this would happen, Sir!' – thus, perhaps, Rutherford, for splitting the atom. 'It was all in a good cause, Sir!' – the discoverers of DDT. 'If I hadn't done it, Sir, somebody else would!' – nuclear fission, shall we say. 'They were going to do it, so I thought I had better do it first' – the biological weapon makers. '*They* made me do it, Sir!' – a general excuse for all servants of all corporate bodies. 'We tried it out, and it *seemed* to work all right!' – the thalidomide tragedy. 'I didn't actually do anything myself; we just talked about it and the other chaps went and did it' – a compendium of justifications for all academic research. 'Well, it does make rather a mess, doesn't it, but everybody wanted to play with our new toy', which covers much of the pollution problem. And so on.

Every such excuse is valid, however essentially infantile. The scientist is not in the front line, pulling triggers and dumping defoliants; by definition, he is a *Back Room Boy,* employed to discover principles and to design devices, not to hurt other people with them. You can't really blame him for what has happened.

True enough; but *if* he had not made that misused discovery, or *if* he had imagined its consequences, or *if* he had not allowed himself to be employed by that evil corporation, then *perhaps* the tragedy would not have taken place. In the complex of social institutions within which we try to make ourselves at home on Earth, the mind and professional expertise of the individual scientist is not a negligible force. The enormous size of the technical community seems to guarantee anonymity and to coun-

* Published in *Impact* (2), 113–22 (1971).

tenance irresponsibility – yet the intellectual leader carries ten thousand of his colleagues with him in a 'breakthrough', and sets ten million humble labourers on a new course of manufacture, commerce and use. We cannot take personal pride in the 'achievements' of our science and technology and simultaneously repudiate responsibility for its failures and abuses. We are either humble workers in the vineyard – or we are indeed the New Men come to make a better world.

The dilemmas of personal responsibility are not new, and the history of Ethics tells plainly that they cannot be resolved. Think of the Inquisition, or of the revolutionary turned executioner, before you put your trust in an ideology, or a pledge of virtue or a Hippocratic Oath. Gospels, social blueprints and other formulae acquire their legalistic interpreters, until the call for peace becomes a war cry and the stake is an instrument of mercy. I see no salvation in resolves or resolutions, however aptly phrased and apparently benevolent. To accept them without reservation is essentially to abjure responsibility; it is the abandonment of judgement and a flight from rationality itself. An innocuous pledge can do no good until it is called into question, when precisely the issue of interpretation, in particular unforeseen circumstances, must be faced. It is then that we must rely again upon our spiritual and intellectual resources – the will to do good, the imagination of suffering, the rational calculation of the consequences of action or inaction.

Social responsibility in science rests therefore upon the way in which scientists are made. Whether or not we have the inborn talents for success in research, we are moulded by upbringing and education. But social responsibility is not a subject to be learnt from a course of lectures (Tu. 10, Fri. 10, Room G.44; Dr Piravetz: Practicals, Sun. 2–5, Trafalgar Square!). It is not something one can practise ostentatiously, as an example to the young ('I think I shall go out and do some social responsibility in science this afternoon. Anyone coming with me? We could count it as part of your optional Field Work'). It is an attitude of mind, a sensibility of the spirit implicit in an educational system, in personal relations, in institutional policies.

What is missing from the education of present generations of scientists? First – they lack general education. They go out into their corporate laboratories as learned ignoramuses knowing all about nuclear magnetic resonance, or the physiological function of adenosine triphosphate, but without any grasp of history, of philosophy, of political thought or of economics. The microbiologists scorn ecology, the nuclear physicists know nothing of warfare, the mechanical engineers are totally ignorant of the physiology of respiration – and so on. In the mad rush to produce completely trained specialists on the cheap, we assume that they will somehow pick up the rest of the knowledge they need. How? From newspapers, from bar room gossip, from TV programmes? We instil into our technologists the highest professional standards, so that they are rightly suspicious of the claims of anyone but an expert in their own field – and then we entrust them with tasks that demand expert judgement over many fields. How can the aircraft engine designer take responsible decisions about noise? That is not his field. There are 'acousticians', aren't there, to deal with that? We need science 'generalists' – not just to run big business or to go into politics, but to do science itself. Our specialized courses of study – pure physics, pure chemistry, pure biochemistry, even pure medicine and pure engineering – are absurd and nonsensical as a training for active life, whether in research itself or in technical development. The assumption is quite false that the clever schoolboy, drawn through the successive dies of primary school, 11-plus, secondary school, O-level, A-level, university, B.Sc. Ph.D., etc., like a billet of steel drawn down into a mile of piano wire, can acquire incidentally all the detailed information he needs outside of such a speciality. The interdisciplinary, technological, historical and economic aspects of our academic 'disciplines' must be *taught* – positively, wisely, expensively and at length.

We must also deal with the political and moral questions associated with science. A course in civics in 5B is not enough – nor is any amount of doctrinaire Marxism. The industrial revolution was a *political* event, with deep moral consequences, not merely a triumph of technique. How was it possible, we must ask, for the rivers and winds of our beautiful land to be so bemired? What forces of greed, or ambition, drove on the mill owners and railway builders – and by what other forces were their monstrous pretensions curbed? How do political and economic forces exert their leverage, and to what extent is science harnessed to anti-human juggernauts? What should be the goals of research? Why should science be regarded as a generally beneficial activity? Should scientists be responsible for their discoveries? Is it feasible to plan research? By what criteria could one begin to judge whether it is better to feed a multitude than to ride like a witch to the Moon? The questions are countless, the moral and political considerations conflicting, tangled and subtle. We need immense drafts of discussion and argument on all such topics, in the seminar room, round the coffee table, in debates, study groups and tutorials, as strengthening medicines for the grown-up world in which these issues will assume reality. The student protest movement can certainly not settle, by brickbats and obscene invective, the dilemma of the county council in choosing between a new hospital, a better bus service, or an improved sewage works. The problems of social responsibility always arise in highly technical contexts, where expert opinion is pitted against expert opinion in the language of cost-effectiveness, budget deficits, manpower projections, ton-miles per litre, perceived decibels and other jargon. We must somehow sensitize these earnest owlish experts to think of people, of pain, of freedom, and of beauty. When they are middle-aged, and grey-haired, and with the power at last to make such decisions, it will be too late. Sophistry and calculation will have taken over – the spiritual lobes in their ponderous noddles will have died for sheer lack of exercise.

Not school or university, but the experience of childhood may decide another important issue – corruptibility. Of all the despicable traits of the modern academic, nothing is worse than the hypocrisy with which he will receive money for research from organizations whose prime ends he inwardly detests. The biochemist who works on a project paid for by a military agency, knowing that it is somehow connected with biological warfare but pretending to himself and others that it is all good clean pure science, has sold himself to the Devil; it is really a pleasure to see such characters now getting hurt. I don't mean that scientific work for military ends is itself immoral. Pacificism is an admirable doctrine for life, and can only be respected, but so also can the principles and practice of self-defence against crime and violence. In civilized countries we support our armed forces, though we may deplore their need and regret their expense. To be a soldier may not be everybody's choice, in peace or even in war, but it is not a dishonourable profession. What must surely be said, however, is that the scientist who is contracted to do research on behalf of a ministry of defence has become a technical soldier. If his country goes to war, then he too is pulling the trigger and dropping the bomb. He may, as a good patriot, feel that he is doing right; that is a question worth quiet debate in each instance. But he cannot take the money and still claim the privileges of a civilian. In civil wars, people get shot for much less than that. Hypocrisy and opportunism are learnt at the mother's knee; the best we can do is to shame those who practise them.

We must learn, in fact, to clarify the essential conflicts of *loyalty* in the practice of science. For the old-fashioned academic it was easy – his loyalty was given, first to his subject, then, in principle to humanity, although a good rousing war cry would quickly awaken his patriotism. The only really important thing to do in life was to add another few papers to the literature of microteleonomy or macroscopology, which was itself, of course, of profound cultural significance because it was 'there'. It was not quite such a noble calling as it was sometimes represented, but once one had got oneself established one certainly knew who, what, and where one was. Nowadays we have to think of our employers, who need to make profits or war, of our country, which ought to make peace, of our professional association, which ought to make up its mind, of our students, who need to make good, and even of our families, who ought to be made to behave. I don't pretend to know my way through these thickets of obligations; one just tries to balance them up as best one can. But part of our moral education in responsibility must be an attempt to analyse these loyalties, lining them up in order of priority.

This is important, because mistaken assumptions of primacy of allegiance can completely spoil a good argument about social responsibility. The SRSs, for example, have been telling the CBW boys that they are all monsters because they are using scientific knowledge 'which is for the good of mankind' in an evil cause – war. Now there is no clause in the Social Contract, the Talmud, the Koran, or the Analects of Confucius stating that scientific knowledge is for the good of mankind, or even that scientists must be cosmopolitan pacifist internationalists. To this the CBW gang have every right to reply that they love their country, that they don't propose to see their sisters die horribly of enemy anthrax, and they will go right on with their patriotic (if unpleasant) duty. The real loyalty, of both parties, should be to rationality and to general humanity – as far as one can make out, biological warfare is the most idiotic form of weapon that could possibly be imagined, suited only to the mutual extermination of suicidal maniacs.

Yes, in the end, it comes down to a passion of spirit but a very cool rationality of intellect. Teaching responsibility in science is also teaching science – the correct appraisal of situations and forces, the use of every bit of knowledge available, readiness to observe, interpret, experiment and theorize. Provided that we give maximum weight to genuine social needs and aspirations – that is, we treat people as people, not as abstract mouths, reproductory organs, sources of exertion or transportable packages – then we must use our heads to the full. Nothing will more quickly discredit the SRS movement than the impatient resort to the impatient slogan or doctrinaire catch-cry. Scientists who thus discard their intellectual powers, their real claim to skill and authority, betray their profession and reduce themselves to the level of John Doe and Richard Roe. Responsibility in science is the use of one's scientific talents, not the pompous authority of one's name on a petition or a vulgar display of naïve political prejudices.

Individual responsibility may demand personal sacrifice. It may mean a willingness to stand up and be counted, to earn public abuse – even the loss of office and employment. 'The blood of the martyrs is the seed of the Church', they used to say. The great communion of science is not unlike a religion, or a church, in our modern society. The doctrines of observational accuracy, rational theory and experimental verification shall be our Trinity, with the President of the Royal Society as our Pope and the Nobel Laureates as our patron saints. With the Science Research Council as a College of Cardinals, with laboratory directors as abbots, with the great accelerators and radio telescopes as our cathedrals, the model is complete. But alas, we have no *martyrs*. Since that equivocal episode of poor old Galileo, it has been a wonderful success story – a primitive sect waxing mighty until made one with the State. Without conflict, without blood, without the opposition of the temporal to the spiritual power, we have been incorporated in the Establishment. The issues of 'science policy' are mere manoeuvrings by our bishops amongst the other great lords of the power élite. They are not allowed to touch on fundamentals – intellectual independence, academic freedom, the right to speak out and the duty to be silent – for these have been sold for money, men and machines at every point of potential conflict.

I do not say that the State is an abomination, nor that science is being used exclusively for wicked ends by power-mad tycoons and politicians. In our democratic societies, there are means of resolving conflicts of goals that do not call for violence and ultimate martyrdom. Yet one must note that such conflicts are latent in the confrontations of transcendental knowledge by earthly economic and political forces. Science under Nazism and Stalinism is the example; how few were those who stood against these perversions of society, in the name of science and truth. Solzhenitsyn's Nobel Prize is for literature, but his picture of the imprisoned science of *The First Circle* is a masterpiece, not only of the imagination but of sober sociology. The scientific achievements of the prisoners are hampered by folly, suspicion and incompetent management, not primarily by a refusal to work for the system that employed them. The myths by which these clever intellectuals lived did not command them to invite martyrdom in place of cooperation; the few who accept this sacrifice do so almost out of a spirit of self-destruction and solidarity with the oppressed rather than heroically in the service of a greater cause. The full story of this episode has not yet been written – indeed, the events themselves have not rolled on to a significant conclusion – so that we do not know to what extent the myth has taken root and grown strong amongst those who suffer and have suffered. Perhaps we shall witness, at last, a community of scientists who know that their genuine responsibility is to stand against lies, hypocrisy, cruelty and folly, not just to acquiesce prudently to the improper demands of state power.

The problems of individual responsibility in science are not trivial. They are not solved by wild movements of protest by passionate students. The responsibility has to be exercised by scientists themselves – that is, by persons of sufficient maturity and experience to speak and act as experts in extremely complicated technical situations. This very maturity and experience can be acquired only by long years of purely professional activity as assistants and subordinates in large-scale enterprises, where the passions are sapped, the moral insights are dulled. All that we can do, I believe, is to sensitize and arm their consciences, sharpen their understanding of the world and its ways and exercise their moral faculties, in youth, so that they may comprehend the issues, and have the courage to face opposition, when the real battles have to be fought.

But let us not despair. The historical answer to the tyranny and irresponsibility of individuals and institutions is not merely personal martyrdom, cunning accommodation, or the preaching of unheeded sermons. The wisdom of society creates the countervailing corporate power – Parliament to curb the King, the courts to curb banditry, trade unions to curb the exploitation of labour, and so on. The 'balance of power' model, the adversary principle, the peculiar institution of 'Her Majesty's Loyal Opposition', are examples of a technique that could well be copied. If we have bodies of scientists dedicated to essentially irresponsible and selfish ends, of profit or power, then we must invent 'antibodies' to neutralize them.

This is the theory of the Food and Drugs Administration, and other regulative agencies of the United States Government. Despite many shortcomings, they do, in fact, perform their allotted functions. In principle, if imperfectly in practice, they oppose the collective powers of teams of scientists to the teams in the manufacturing corporations, thus acting as deliberately conscientious elements on behalf of society.

Within the framework of scientific methodology, this is the correct procedure. The task of the scientific innovator is to persuade the other members of the scholarly community that he has made a valid contribution. The task of the others is to oppose – not out of mere conservatism and prejudice, but as informed critics; unwilling to be convinced by mere assertions. The progress of knowledge is dependent upon such debates – not elevated into personal controversies, and always tempered by tolerance and scepticism on both sides. The apparently certain tone of each particular scientific paper belies the underlying uncertainty of the issues; it is the duty of each participant to use his persuasive powers to the full, just as the duty of the advocate is to make the best of his case before the court.

Scientific responsibility in social issues therefore demands such debate, whether in the comparative privacy of the learned journals or in the forum of the Press or Parliament. In most cases, there is no absolute truth to be determined: DDT is both a blessing and a scourge; motor vehicles are convenient, but noisy and dirty. A balanced report by a single expert commission, however well-intentioned, cannot judge between conflicting opinions and priorities until each party has expressed, to the utmost, its own special viewpoint or interest. What is important is that there should be adequate representation, of a skilful, professional kind, of the interests of the general public, of conservationists, of bird watchers, of the lovers of peace and quiet, of preventive medicine, of anglers, of the League against Cruel Sports, of the preservers of churches and windmills, besides the immediate economic contestants such as industrial corporations, local authorities, public utilities, transport undertakings and so on. It is as much the duty of the state to ensure that this type of expert evidence and advocacy is available as it is in a criminal trial to have proper lawyers for the defence.

But where will such experts come from? Who will normally employ them? Industrial corporations and the various organs of the state know how to provide themselves with scientific consultants; what about the numerous other interested parties, especially the ordinary citizen who is to be assaulted, battered, deafened, poisoned, driven from his home, or otherwise insulted? In theory he too is the ward of a benevolent state, that will develop further agencies for consumer protection, planning, noise abatement, etc. But the conservatism that breeds in such agencies is all too familiar. They continue to plod solemnly back and forth on ancient sentinel duties, and are blind to new dangers. Being in the political domain, they are the targets of political forces, such as those of corporate industry. And who will guard the guardians? Where are the professional experts to countervail the military power, the powers of state monopoly, of national health services, etc.? The US debate about anti-ballistic missiles is a case in point. Although this is not at all a matter of pure science, the public case against the ABM system had to come from a small volunteer group of academic scientists (who are not, indeed, ignorant on the subject) rather than from an engineering organization charged specifically with these interests. What happens in the Soviet Union during such debates is hidden from sight, but beggars the imagination.

Where can institutionalized, licensed dissent and criticism survive, except in our universities? The great issues of academic freedom and independence are more desperate than ever, now that the learned men have real fire to play with. Having demolished the ivory tower in the name of social relevance, we must rebuild it as a watch tower over all matters of technique and social action. We need ABM watchers, and CBW watchers, and pollution watchers, armed not with slogans but with searchlights and telescopes of specialist knowledge.

This is not an easy exercise in social engineering. Your natural ABM watcher is a cinch for a fat contract for secret work on radar systems for the Pentagon; a really good pollution watcher is the ideal consultant on the payroll of an oil company; nobody knows anything about CBW except those who have already been sworn in under the Official Secrets Act. Yet I think it must be attempted. The advantage of the university environment is that it provides a safe professional base for expert criticism; the academic is paid primarily to be a teacher and a scholar, not to provide specific research results for specific corporate bodies. A tenured professor does not have to be personally brave to say what he likes about the government, the local big-wigs and the policies of Generalized Manufacturers Unlimited. True, he may not get his next research contract renewed, but why should he care: it will give him more time to write a book instead of filing innumerable reports and questionnaires. The academic economists, in their splendid running battles with Treasury policy, set us a good example. Chaps like Galbraith and Kaldor move in and out of government circles – and when they are out of favour they don't hesitate to cock devastating snooks at their lords and masters in Washington and Whitehall. How many aeronautical engineers, or biological soldiers, are doing just that? This is not a matter of unusual individual 'responsibility'; it is just the institutionalization of loyal opposition in the academic watch towers. The real problem is how to switch the so-called research interests of many university teachers into such channels, and how to fund them so that they have adequate assistance and equipment without coming under the sway of precisely the corporations and agencies they are meant to watch. This is a field for much more experiment, by bodies like the Ford Foundation and the National Science Foundation in the USA and their equivalents in other nations.

In the end, a country gets the scientists it deserves. A responsible society breeds, trains and fosters responsible scientists. An open market in ideas and political criticism is also open for technical attack and counter attack. A free Press, for example, ready to publish informed articles on scientific matters, may be the essential atmosphere in which kites may be flown as first signals of a storm of controversy. Remember that the single most important act of scientific responsibility in our time was the publication of Rachel Carson's book *The Silent Spring* – surely impossible in a totalitarian society where all scientific questions are automatically solved by appropriately planned groups under the benevolent eyes of the all-wise Party and an omnicompetent Council of Ministers. The problems of technological progress are not, in the end, capable of decision by 'scientific' methods; they are problems of social priority, of aesthetic judgement, of taste, of preference, of material and spiritual standards. The final arbiters must be the common people, as users and abusers of our pretty toys.

Let's not fool ourselves, though. Most 'political' and 'economic' decisions about technology are taken at first hand by politicians, business men, generals and other socially powerful people. Until quite recently, these proud, self-confident and supposedly responsible chaps scarcely deigned to listen to scientific arguments at all; technical change was occurring at much lower levels in society than in their elevated world of political parties and personal profits. Now they have learnt something of the need for expert advisers, and at least some sort of statistical evidence to support their intuitive wisdom. Despite one's suspicions of the closed worlds that some such advisory groups can build about themselves, especially when protected by secrecy, one must encourage this development; for the moment we cannot have too much decent science in the government and in other organs of social action. The Luddite mood of many intellectuals would be disastrous if it communicated itself to the power élite, for it would hamper good, responsible science without putting any obstacles in the way of selfish technocracy. Our civilization has gone too far along the path of bureaucratic and technological sophistication to survive such a repudiation of rationality.

It may be, indeed, that the pendulum of prejudice has swung too far. The public adoration of the scientist, as the sage and saviour, is a thing of the past; now we seem to hear nothing but scorn for his pretensions and hatred of his arrogance. The movement to harness every technical expert to environmental studies or systems engineering, to make him useful, and safe – your friendly neighbourhood boffin – could be as damaging as the older snobbery of pure science for its own sake. Scientific knowledge and social action are not the same thing. The neo-Marxist argument that all science is 'really' determined by social ends is either a vacuous truism or it is dangerous nonsense. Natural philosophy is not entirely for useful ends, despite the technological spin-off. If you try, too short-sightedly, to press it into service for immediate ends, then you will rob later generations of its products. Countercyclically, I feel the need to preserve the collective skills, the expert knowledge and the delicate social organization of the scientific community from the pressures of an ignorant public, a shameless Press, rapacious money-makers and opportunist politicians. A certain aloofness, a slight distance from everyday affairs may be the only way of preserving these islands of sanity in a crazy world, not as refuges but as watch towers and safeguards against far greater evils. That is the paradox: social responsibility in science must not be too concerned about today, for tomorrow also will come.

22

Seeing Through our Seers*

The expert may feel that he has everything to teach – but he has much to learn from others about his specialized view of the world.

The voters of California have spoken on an issue of much greater importance than whether Ronald Reagan or Gerald Ford should be the Republican candidate in the November presidential elections. They have rejected Proposition 15. This proposition, which was placed on the ballot by petition last year, would have set stringent controls on the construction and operation of nuclear power plants in California. If it had passed, it would have brought the nuclear industry there to a grinding halt. Voters in other states would then have been encouraged to support similar propositions in November, and the chances are that the United States would have prohibited nuclear energy with the same fervour that it prohibited the sale of alcohol 50 years ago.

I can think of no precedent for a plebiscite on so technical an issue. From now on, the benefits of technical progress, and the ills that may accompany them, are not necessarily to be foisted on us by profit-seeking capitalists or by an autocratic government. We need no longer drift into a mechanized lifestyle which is not to our liking, or be carried along into a disastrous millennium by the enthusiasm of optimistic innovators. Following the lead of California, we, the people – or, more likely, our political representatives in Parliament or Congress – are to decide on these important issues for ourselves.

But the questions we will have to answer will not be like those of everyday life and everyday politics, where the lifetime experience of every mature person is not altogether unreliable as a guide to action. The new questions involve some of the most abstruse problems of physics, chemistry, biology and economics – questions such as: how many children might be expected to die of leukaemia if the radioactive wastes of a nuclear reactor were to escape into the atmosphere? Or, in some ultimate tragic

* BBC Radio broadcast, published in *The Listener* 95, 794–6 (1976).

scenario, why has it stopped raining in Britain, and where have all the flowers gone?

Who can guide us on such issues? There are plenty of experts, but they offer us divided counsels and refuse to take any responsibility for the consequences of following such advice as we can squeeze from them. The deeper we probe, the more searching our inquiries, the less certain do we become. Whether he is a Californian bank clerk voting on Proposition 15, or our very own Mr Tony Benn planning the exploitation of North Sea Oil, the layman who seeks stable ground on which to base his decisions sinks in a quagmire of doubts and contradictions.

What is wrong with our science? What is wrong with our scientists, that they seem so unhelpful in practical affairs? No, that is a negative way of putting it, for I am not suggesting that they are devoting themselves to unworldly, essentially useless pursuits, and don't care for the needs of ordinary humanity. The goal of science is not to solve a succession of immediate problems but to establish a body of reliable knowledge from which good answers to particular questions may ultimately be derived. A real scientist has not much taste for laying down the law in practical matters. He may be quite inflated with confident belief in his own interpretation of his own tiny corner of the phenomena of nature, but his general style of thought is, inevitably, quizzical and sceptical.

He delights in his ability to find faults in the arguments of his competitors, and devotes a great deal of his time to bolstering his own opinions or proving the other fellow wrong. A fact or a theory that is well-established and uncontroversial is no longer exciting, so he leaves that to dullards or crackpots or devoted teachers, and shifts to more open questions. So, although he may, in fact, know a great deal that is highly relevant to an important practical problem, it gives him little pleasure, and less kudos, merely to expound what he doesn't doubt. And when you do get the benefit of his knowledge, he cannot resist the temptation to exhibit his scholarly integrity by dwelling on all the fascinating uncertainties and discrepancies that he has taught himself never to ignore. The layman, dazzled and disconcerted, retires little the wiser.

Of course, the big boy in government or industry can buy the expertise he needs. He can employ his own scientific advisers and consultants. He may have a whole research laboratory full of experts to answer his questions. And the technological innovators themselves, the manufacturers of monstrous ships or gigantic and intricate computers, or pharmaceutical products, rely upon the skill and ingenuity of thousands of scientists and engineers. If only we could trust them to monitor their own products.

There are, indeed, forms of government where such issues as the development of nuclear power or supersonic aircraft or environmental pollution, or energy resources or population planning, are decided by in-house experts who do not have to present their views for public debate. We call that form of government totalitarian.

It is the wisdom of the pluralistic society to doubt the competence of any authority to choose wisely on behalf of every citizen. It is not so much that they cannot be trusted not to feather their own nests; it is simply that the questions debated within such organizations are seldom correctly posed. They refer far too much to what is technically possible, or technically optimal, rather than what is socially desirable. This is perfectly exemplified by the history of the Concorde project, where the advisory committee seems to have been dominated by engineers and accountants, rather than by potential passengers or by non-passengers residing near large airports.

It would be wrong to suppose that the layman seeking expert advice is something new in human affairs. This is a familiar situation in the medical profession, which has a well-established protocol for the relationship between doctor and patient. The essence of this relationship is that the patient gratifies the doctor with a fee, while the doctor gratifies the patient by telling him what he wants to know – primarily that he will get well, if only he entrusts himself to the prescribed treatment. The prime aim of this relationship is to establish confidence, and this requires that all treacherous doubts should be carefully suppressed. But, since the doctor often has very weak grounds for his opinions, he is at pains to conceal them by insisting, in Alexander Pope's famous dictum, that:

A little learning is a dangerous thing;
Drink deep, or taste not the Pierian spring.

Psychologically speaking, the doctor – patient relationship is a very satisfactory model for dealings between the expert and the layman. Unfortunately, we have come to learn that the advice of the doctor is not always as reliable as he pretends. In a serious case, it is a maxim of worldly wisdom to seek a second opinion – which often turns out to be widely at variance with the first. If you have ever had severe back trouble, you will know what I mean. In other words, the advice given by the professional expert, in the confident style of the doctor, is not really less controversial than that of the scientist. Because he is in the advice-giving profession, he presents his opinion in such a way as to encourage confidence in his own views, rather than emphasizing the errors of his rivals.

Don't get me wrong. I am not attacking the extraordinary achievements of modern medical science and technology. I just wanted to point out the dangers inherent in the social role of the professional technical adviser. The recent increase in malpractice suits against physicians and surgeons is not a symptom of social health, but it does indicate the growth of scepticism in the laity, and a proper disrespect for the mystification in which a guild of professional experts may attempt to enshroud themselves to avoid unwelcome criticism.

The impotence of the layman in the face of technical expertise is not confined to the general public. Thoughtful politicians and industrial managers realise that, despite their access to greater resources of organized brain power, they still have the same problem. They begin to see themselves as much the victims or prisoners of the technologists as those more humble persons who must simply suffer for mistakes made on high. Governments in many countries are busily inventing new administrative devices to deal with this problem. The usual technique is to create monitoring or regulatory agencies where teams of experts are employed to supervise technological innovations.

In the tradition of the constitutional separation of powers, the United States Congress has created an Office of Technology Assessment to advise it on the pros and cons of new technology. The United States Federal Government seeks opinions on a wide variety of issues from the National Research Council – a companion body to the National Academy of Sciences – where innumerable experts are convened in committee to deal with specific questions of science and technique. I am forced to draw these examples from the United States, because I don't think we yet have anything comparable here in Britain. It is only recently that the House of Commons Select Committee on Science and Technology has begun to ask questions as tough as those of its counterparts in Washington and to provide serious countervailing power to the in-house expertise of the Civil Service.

The most reactionary proposal, despite its good intentions, is to establish a Science Court, which would provide authoritative decisions on technical problems. The judges would not themselves be experts on the matter to be decided but would listen to arguments from both sides and then send down a judgement which would henceforth be regarded as a fixed point of certain knowledge. Being wise and learned men, such as Nobel Laureates, the results of their deliberations would be of sufficient weight to inspire full confidence in the decision.

I am quite sure that the open debate that would take place in such a court would be of the greatest interest and value. There is nothing like an organized advocacy process to get at all the evidence relevant to a complicated issue. And some of our scientific and technical colleagues get away with far too much because they never have to face skilful public cross-examination of their views.

But what would be the value of that final judgement? Look at the current controversy concerning the effects of aerosol propellants on the ozone layer in the stratosphere. A few months ago, it seemed to have been clearly proved that the consequences would be disastrous, and the innocent practice of deodorizing oneself out of a pressurized can of sweet-smelling fluid must be stopped at once before mankind gets sunburned to a sad end. But then a new calculation suggested that the key material was something called chlorine nitrate – an impossible compound, according to my high-school chemistry – which would remove all the poisonous substances and leave the ozone in its natural state. Well and good, but then I noticed the other day that this process might not be as efficient as was claimed, so that we could be back in trouble after all. In other words, our resplendent court could not honestly come to any better decision than that more research was needed on the subject – the standard academic response.

The fundamental defect in all such proposals is the desire to achieve a feeling of certainty based upon unimpeachable authority. Yet even the unanimous positive decision of such a court might not be as politically effective as its proponents imagine. A case in point is the fluoridation of water supplies to prevent dental caries. I think it is fair to say that the overwhelming consensus of well-informed experts favours this mild practice. I can scarcely believe that a well-constituted Science Court would shilly-shally on such a straightforward issue. Yet mass distrust of expert opinion is now so much an element of political life in democratic societies that fluoridation has been successfully opposed in many communities. I do not believe that the doubts on which the fanatical opponents of fluoridation play are those of genuine scientific scepticism, but reflect the over-authoritarian attitude of the doctors in the past. With little obvious risk to life and limb, we can vote against them and show our disrespect. To quote Pope once more:

> Be not the first by whom the new are tried,
> Nor yet the last to lay the old aside.

I am not, myself, going to offer another specific remedy for this ill of the body politic, but my diagnosis of the underlying disease is that it arises from the mistrust between expert and layman, and the fundamental differences in their perception of, and approach to, technical problems.

This 'cognitive dissonance', as the sociologists call it, is so severe as to hinder mutual comprehension. The institutional models to which I have referred are psychologically based on the adversary principle. The truth, or the optimum decision, will be arrived at by confrontation and conflict. Within the scientific world, where the advocates of opposing views share a common technical language, this is a very powerful and effective principle. But in the relationship between the layman and the expert, we must, surely, be seeking cooperation rather than confrontation. To achieve the necessary cognitive harmony, we must use the psychic forces that act on people when they try to combine to achieve a common purpose. We need to exploit the educative effects of working together.

An old friend of mine, a theoretical physicist, described recently his experience of working with experts in other disciplines – chemists, mathematicians, biologists – in a team that was to report on a complex technical problem. What he learnt was humbling – that they, too, had valid attitudes and intellectual techniques which were difficult at first to understand, but which were worth mastering and sharing. The way to grasp a complex technical issue with social repercussions is to bring experts and laymen together within a decision-making machinery: on committees, on working parties, in political organizations, voluntary movements, bureaucratic systems, etc. Instead of standing off at a distance and firing shots at one another, they must be forced to talk so intensively together that they begin to understand one another. In his arrogance, the expert may feel that he has everything to teach – but he has much to learn concerning the attitude of others towards his specialized view of the world.

The proper antidote to the poison of 'technocracy' is something like 'participatory democracy'. Of course, I don't believe that all the voters in California could have been made wise about nuclear power by such processes. The fundamental problems of the unequal distribution of education, experience and responsibility amongst the voters of a democratic society are ever with us. But I do not believe that we should regard the underlying issues of technological development as inaccessible to the layman who is prepared to take the trouble to inform himself about them.

In this process of education and self-education, the popularizing journalist plays a very important part. His job is to bridge the cognitive gap from the side of the layman. In the case of the nuclear power debate, for example, the Californian voter, faced with Proposition 15, had much more to go upon than the bold assertions of the experts on either side. In a whole range of journals, from popular daily newspapers, through the weekly magazines, to quasi-technical organs, he could find a variety of articles summarizing the issues in reasonably clear language. The broadcasting media, especially television, provide more vivid interpretations for the general public. The popularization of science in the mass media is thus of immense social value. The good technical journalist can acquaint himself with the essential expert opinions and draw attention to the most significant features of a complex argument. If he really knows his stuff, he can distinguish between genuine expertise and the bogus authority claimed by those who produce – again in Pope's devastating lines –

> Such laboured nothings, in so strange a style,
> [that] amaze th'unlearn'd, and make the learned smile.

There are dangers, of course, in relying upon the journalist. Unlike the doctor, who feeds on the favour of those who buy his words, and thus plays to their desire for safety and certainty, he tends to fall into the opposite temptation of discovering disquieting features and 'viewing with alarm'. Competition for public attention spoils his judgement. What he tells us is interesting, often highly instructive, but not altogether to be trusted in our search for understanding and decision.

But the essence of my argument is that the mystery-mongering of the doctors is the wrong approach to problems that will grow more complex. Traditional medical practice has been concerned with the prescription of remedies for diseases that are already manifest; the essence of preventive medicine is to educate people into an understanding of their own bodily processes so that they can take their own responsibility for leading healthy lives. To deal with the ills of technological progess, we need sufficient general awareness to prevent follies and failures before their consequences have become so disastrous that all we can do is to call in the doctors and submit ourselves to a drastic cure.

23

Is War good for Physics?*

The Institute of Physics is concerned with all aspects of physics. War interacts strongly with physics, and is therefore a legitimate topic for discussion in *Physics Bulletin*.

Physics is certainly very good for making war. The only question is whether nuclear weapons may have become so destructive as to defeat war itself. But is war good for physics?

Engineering undoubtedly benefits enormously. Aeronautics and electronics leapt forward decades between 1939 and 1945, and rode high on the crest of the Cold War. This arms race continues. From weapons technology physics gains innumerable techniques. Radio astronomy and magnetic resonance grew from radar microwave practice. Spin-off into mathematical methods, computers, materials, etc., can scarcely be evaluated.

Has basic physics gained conceptually from warfare? Galileo learnt much from artillery ballistics, and Rumford observed the heat produced from the boring of cannon; it is hard to think of more modern equivalents. It is not recorded whether the time spent by some leading theoretical physicists in planning electronic battlefields has contributed specifically to their attack on quarks and quasars.

The great boon has been more money for research. Something between 20 and 40 per cent of the total R & D expenditure of an advanced industrial nation is charged to defence. Since physicists are particularly effective in war research, this buys an immense amount of physics. Relevant fields such as nuclear physics, oceanography and quantum optics have been pushed forward rapidly. Deliberately or indirectly, lavish funds have also been provided for many fields which seem intellectually ornamental rather than immediately useful.

Money means jobs. With employment not so overfull as it once was, war research is keeping the physics profession from disaster. But not all

* Leader in *Physics Bulletin*, 24, 583 (1973).

those who, ten years ago, opted for this sort of work have managed to retain their jobs; it might have proved safer to choose high school teaching after all.

It is a mistake, of course, to find oneself in the firing line. The survivors in Dacca, or Nsukka, can vouch for that. Scientific apparatus does not survive well against bombs and bullets, nor do physicists. On the other hand, there were advantages in being qualified to devise new antipersonnel grenades in an air-conditioned laboratory in California, instead of being drafted to dirt and danger in the Mekong delta. Shooting war is seldom so bad for physicists as for most other people. But if you happen to be a Russian Jew you can't win; whether or not you have been patriotically contributing to the defence of your country, you can always be accused of having been involved in secret defence work and denied an exit visa.

The above remarks do not exhaust the subject, but may suggest, for those with a taste for such exercises, the categories for a more detailed cost–benefit analysis of this interesting social phenomenon.

24

Professional Scientists and their Impotence*

In an era of industrialized research, there is need for organizations to protect the personal autonomy of the research worker.

Joseph Ben-David's analysis is of such richness and depth that only a long book could do it justice. In particular, his conclusions concerning the inadequacy of the scientific community as the guide to its own destiny – what physicists would call the 'bootstrap model' – are all too necessary and true. I should like to pursue a few points which he raises even though I might seem to develop them in a slightly different direction.

Professor Ben-David and I start from the same thesis – namely, that scientific research is always associated with a relatively well-integrated community of scientists. To my mind, the existence of an intellectual community is itself a pre-condition of science. In the attempt to construct a satisfactory methodology or philosophy of science, one encounters at every turn the social relationships between scientists. The 'objectivity' of experimental evidence is its interpersonal invariance; the logical necessity of theory depends upon the rationality of language and mutual social consciousness. The goal of science is to create a body of knowledge which is publicly agreed by all competent authorities. 'Consensibility' is the criterion, and 'consensuality' the ultimate purpose of all valid scientific research.

The 'scientific revolution' dates from the seventeenth century, when the machinery for the exchange of research results and critical comments was brought into being in Western Europe. Without such machinery, the activity of would-be scientists (witness the alchemists) would not merely have seemed socially deviant and subversive; it just would not have been science at all. It is beside the point that this machinery – especially the learned journals – was constructed, fostered or protected by organizations such as the Royal Society and the national academies, which tried to play

* Letter to *Minerva* 11, 133–7 (1973) commenting on an article by Joseph Ben-David 'The Profession of Science and its Powers', *Minerva* 10, 362–83 (1972).

a more deliberate part in the legitimation, management or patronage of research. The significant feature was the opening of a new communication network linking more closely the members of the international scholarly community.

For the individual scientist, therefore, membership in, or access to, this network is the *sine qua non* of his calling. Before everything else, he needs to communicate his discoveries to his peers, and to be acquainted with their comments, criticism and subsequent contributions. He may not in the least care whether the whole body of science is socially esteemed, or whether the public at large needs the assurance of a caucus of official pundits before it will accept his findings. Indeed, he might enjoy a perverse pride in the unorthodoxy and outlandishness of his final conclusions, so long as he has been given an adequate hearing by those few of his contemporaries whose judgement he has come to respect. For the past three centuries, to be a *savant* or 'scientist' has been to give intellectual allegiance to that particular human group – a group which has transcended the frontiers of race, religion, nation or province. It is precisely the metaphor of the 'invisible college' which has sustained him, and made his labours so fruitful.

Until recent years, the scientific community as a whole has had no more coherence or corporate existence than the bare minimum needed to maintain this communication system and to preserve the sense of spiritual fellowship amongst its members. Indeed, if natural philosophy and natural history had remained, like fox-hunting, the obsessive pastime of gentlemen of independent means, this *collegium* could still have preserved its invisibility. A great deal of valuable research was done by just such gentlemen – especially the clergy of the Church of England – throughout the eighteenth and nineteenth centuries, without the benefit of tight professional organizations. In France they thought they had ordered these matters better, their university teachers being civil servants; but that was the French way of life and not the inevitable pattern for the advancement of science in that era.

Unfortunately, amateur science has not satisfied all our aspirations and needs. Professionalism – by which I mean no more than being paid to do research – was inevitable. Not all those with a scientific bent have been born rich, or have had the foresight to provide themselves with an undemanding occupation such as a rural benefice. If appropriate support for scientific talent can be found, then it is easy to argue that the world will be the better thereby.

Nevertheless, the *savant* is liable to suffer a serious disadvantage when he loses his amateur status: he may have to please his paymaster. The general question whether 'Science' should be supported by 'Society' is beyond our comprehension: it juggles with numerical abstractions, such as manpower statistics and research budgets. One must bring it down to the human level of Dr X being granted a fellowship to investigate topic A, and then ask what is being offered and what is being given.

It is not at all like a contract for 'professional services' as if with an architect or dentist; nor is it simply the employment of a technical expert like an engine driver or bookkeeper. The nearest analogy is with other forms of creative work. The artist or novelist supports himself by the sale of his products *after* they have been created. Similarly, in the German academic career described by Professor Ben-David, the *Privatdozent* laboured at his research for many years before he was 'rewarded' with a professorship. He was not obliged to work on any particular problem (though there were circumstances such as his training which did restrict this freedom) and he took upon himself the risk that his research might be fruitless. The cost of unfulfilled scientific hopes was borne by the individual, in a disappointed life and wasted personal resources, rather than by the universities and academies which provided the rewards. That is the message of Max Weber's moving essay. The corresponding phenomena in the competitive worlds of literature, the arts, the stage, etc., are all too familiar.

And if we ask why provision is made, by private foundations or by the state, through academies and universities, for these attractive posts and titles, it is difficult to come up with answers which differ significantly from those that might be given to the question why Lorenzo de Medici was a great patron of the arts, or why a succession of princes could be persuaded to support Erasmus of Rotterdam in his wanderings. There are superficial practical reasons (was not Leonardo officially employed as a military engineer?) but the simple answer is that the patronage of learned men adds lustre to the state and to its ruler. That lustre is all the brighter if the support is given generously, without strings. To ask of an academy of mathematicians and astronomers whether they are giving adequate returns by way of nautical almanacs and the design of ordinance is like asking of the windows of Chartres cathedral whether they keep out the rain. The gain in this business is in glory, not in discounted cash flow or real estate.

Unfortunately, this pattern of professionalism is unstable. It is difficult to devise a graded succession of rewards for outstanding scholars without at the same time supporting a good deal of mediocre research by those who inhabit the lower ranks of the hierarchy. The patron can then scarcely resist calling the tune, and insisting that this work, at least, shall come under more inspired direction.

The scientist has no product which he can sell in the open market, to protect himself against such exploitation. To imagine his situation, suppose that there were no copyright laws, and that original works of art were not more prized than manufactured copies. The modern doctrine of the right and duty of the creative artist or writer to 'express himself' would be completely undermined, and we should return to the ways of the Middle Ages, when the artist was no more than a 'craftsman', commissioned to produce appropriate objects for the use of rich individuals or corporations. By a quirk of fate, science has followed this trend in reverse. From an epoch of individualistic self-expression, we have moved to an era where the 'scientific worker' is usually a craftsman employed on a specific commission.

It is unnecessary to cite formal evidence for this historical phenomenon. Every advertisement for a scientific or academic post under the rank of full professor spells out the nature of the research to be carried out in terms as unambiguous as for a post as a bank manager or as a travelling salesman in ladies' garments. Modern science is not a 'profession' in the Victorian sense, like medicine, architecture, or law, with a high proportion of 'self-employed' private practitioners; it is highly industrialized into a small number of factories or firms where individuals work in teams using expensive capital equipment which is not their own property. When talking or thinking about academic freedom, or the intellectual autonomy of science, we should look at this reality, rather than at the melodramatic antics of the mandarins as they struggle with the war-lords. Beneath this aristocratic layer, we are all becoming skilful household slaves.

The clock cannot be turned back. The goal of consensual public knowledge can be approached far more rapidly by the team work and expensive instrumentation of 'big science' than by cottage industry research. Patronage for charismatic scholarly ability used to be provided by academic or civil service sinecures; it cannot survive the scrutiny of the accountant or the rationalization of the management consultant. The whole scale of operations has become too large, and must be justified by more immediate returns such as useful technology. Bureaucratization and industrialization of the scientific community are inevitable if we are to make any sense of the proliferation of scientific institutions and people.

How, then, should research be managed, controlled or planned? The most serious error is to put these powers into the hands of the organs of the intellectual community – the learned societies, the national academies, etc. The forces at work in the intellectual sphere – the logical necessity of hypothesis A, the insights of Dr W, the credibility of Professor Q – are not commensurate with those in the political world, where money and institutional rank must do the talking. To conflate these forces, without recognition of their dimensional incompatibility, is bad for scientific knowledge and bad for politics. The separation of the executive and judicial functions of the state is the most appropriate model. It is significant that Stalinist totalitarianism, which overruled all barriers of personal justice, was also fertile ground for the scientific plague of Lysenkoism, expressed through the combined 'intellectual' and 'management' function of the Soviet Academy of Science.

This is no more than conventional liberalism. But we might take the argument a little further, and ask whether adequate countervailing power now exists to support the individual scientist against the organizations which employ him. The learned societies, with their commitment to the intellectual sphere, are traditionally dominated by the academic individualism of charismatic leaders, who themselves enjoy great freedom as the holders of sinecures, or who wield personal authority over many subordinates. It is in their interest (and, as I have remarked, not merely for expedient reasons) to preserve the intellectual sphere from the corruptions of 'politics', which they can, in any case, play out on other fields. We may also observe some flat-footed trade unions growing up, with an interest only in pay, working hours, fair employment practices, etc. Unfortunately the policies and constitution of these organizations strongly favour further bureaucratization of science on behalf of the mass of members who seek security as advanced technicians rather than the risks and uncertainties of open-ended research.

For the health of the scientific community, and for the continued efficacy of the scientific 'method', there is need for some institutions which will express and protect the individuality of the scientist as a social being, and his creativity as a social factor. Numerous issues concerning scientific education, freedom of research and publication, the ethics of sponsored research, and other themes of 'social responsibility' need to be talked out, and carried to the level of effective action. Such themes are now finding their way into the agendas of the learned societies, where they cause much embarrassment and conflict because of their political overtones. But there is no reason why they should not become the special business of a general organization of scientists – for example, the British and American Associations for the Advancement of Science – covering all disciplines.

Professor Ben-David reminds us that a component feature of a professional occupation is 'a limitation on the contractual obligations of the professional towards his client or employer'. In my opinion, this feature is inadequately developed in the profession of science as it now exists. I do not pretend to know the precise form which these limitations should take, nor the subtle checks and balances, precedents and traditions, corruptions, expediencies and ethical principles in which they would be embodied in practice. But that would be a more suitable goal for action by the scientific community, in defence of its own autonomy, than some delicious power to distribute a large slice of the national income for the relief of its own members.

25

Can Scientific Knowledge be an Economic Category?*

Knowledge does not conform to the axioms of arithmetic.

What does economics tell the student of 'the sciences of science'? It is not unreasonable to conclude from reading this book that the whole subject is still shrouded in obscurity. This is essentially what the authors themselves tell us, for it seems that none of the really interesting questions about the economics of research and development can yet be satisfactorily answered. The general statistics of the resources devoted to science and technology in various countries can be studied, but neither the reasons for their variability nor the consequences of national differences have been interpreted. No conclusion seems to have been reached to any of the old arguments concerning the economic effects of the patent system, the value of basic research to individual industrial firms, the effects of monopoly on research and development, the features favouring energetic innovation by a particular firm or in a particular industry, the circumstances governing the rate of diffusion of a technical innovation, the measurement of the social benefits of research, or the contribution of science, research and technological innovation to general industrial growth and national development. The observation that an innovation diffuses into industrial or domestic use according to an 'S-shape' curve is scarcely surprising: how else can a monotonic function rise from 0 per cent at the beginning to something approaching 100 per cent a few years later, without kinks or discontinuities? The failure to discover a significant factor for 'advances in knowledge' in various measured contributions to economic growth does not shake our common-sense belief in technical progress; it merely proves that the statistical data are inappropriate for this task.

One major question is only briefly mentioned: what are the economic consequences of highly concentrated, sophisticated and extravagant research on the technology of war? Since defence budgets cover something

* From a review of *The Economics of Research & Technology* by K. Norris and J. Vaizey (Allen & Unwin, 1973), published in *Minerva* 10, 384–8 (1972).

between 20 and 40 per cent of all research and development expenditure in advanced industrial states, this factor deserves more than several pages of vague remarks concerning the aerospace industry. For example, the relative strengths of research laboratory 'push' and armed forces 'pull' on innovation in military technology might have been discussed, or the problems of balancing incentives against waste in 'cost-plus' research contracts. But these questions would probably not have received convincing answers, either.

It must be emphasized that Dr Norris and Professor Vaizey are themselves apologetic about the inconclusiveness of the work they describe. The question we must begin to ask is whether this is inherent in the whole enterprise. We must begin to doubt whether the technical knowledge which is discovered by research or learnt by experience can be treated satisfactorily as a category of economics at all.

Economics is concerned with the measurement and rational allocation of the material resources of society to the achievement of whatever goals that society may have chosen. Like physics, it is essentially a quantitative science, where verbal or pictorial representations are no substitute for mathematical proofs. The fact that research is nowadays a paid profession, the costs of which appear in corporate balance sheets and national budgets, does not imply that its product must be a staple commodity, the value of which can be entered on the other side of the ledger. Although scientific knowledge cannot be dismissed as an epiphenomenon, like chamber music, without relevance to economic activity, there are formidable objections to any attempt at quantifying it in monetary terms. These objections are often voiced in a vague way by economists, but their full significance as obstacles to mathematical model building are not always appreciated.

Suppose, for example, that we had constructed a theoretical model in which a particular piece of knowledge is assigned the value *x;* what algebraic operations would be valid for this variable? A moment's reflection reveals that nothing like ordinary arithmetic – which is, of course, the algebra of money – can be employed. This follows at once from certain very simple characteristics of knowledge:

Non-equivalence Economic theory depends upon the identity of objects such as motor cars from the same assembly line and their equivalence in exchange with so many barrels of oil, etc. But each piece of knowledge is unique, and cannot be regarded as equivalent to any other piece with which it might be exchanged. To know the electrical resistivity of liquid sodium is not the same as knowing the corresponding quantity for copper:

fast breeder reactors are cooled with liquid sodium, not with liquid copper. In the algebra of knowledge, we can never write $x = y$.

Non-additiveness The intrinsic value of a piece of knowledge depends on what other knowledge goes with it. The Atomic Energy Authority is willing to spend a lot to find out about the electrical properties of liquid sodium because it already knows that this metal does not absorb neutrons heavily; otherwise, this particular piece of knowledge would be of mere academic interest. In the arithmetic of research $2 + 2 = 0$, or $2 + 2 = 100$, according to circumstances.

Irreversibility Once something is known, it cannot be 'unknown'. Thus, the acquisition of knowledge is an irreversibly cumulative process, to which the operation of subtraction is not applicable. This bank account can never be in the red.

Indivisibility A piece of information, once known, cannot be separated into its parts and shared amongst various individuals like a piece of cake. In this realm, half a loaf is no loaf at all.

Infinite-multiplicability For practically no cost, a piece of information can be reproduced, multiplied, and broadcast in any quantity. How is its 'value' changed by this process? If x is a secret, than $2x$ is less than x, and $100\,000x$ is effectively zero.

Until these objections are shown to be groundless, or until some genius turns up with a completely new calculus in which the operations on 'the value of knowledge' do not have to satisfy any of the ordinary rules of arithmetic, we must surely be sceptical about any claim to have weighed this large but imponderable agent in human affairs.

In desperation, we might attempt to set up a statistical model, in which large numbers of bits of information flow out of research laboratories at an average cost and generate statistically smoothed average benefits. But the variance in the intrinsic value of pieces of scientific knowledge is enormous. A theory from which one could calculate the mechanical strength of any alloy would be worth infinitely more than the sum of all such pieces of information about existing alloys, since it could also predict the properties of future materials not yet tested. Statistical analysis is meaningless when applied to aggregates, the individual properties of which range so widely; the average price of a mixture of diamonds and glass beads is not a useful economic index.

A more modest theoretical programme is implicit in most of the work reported in this book – to treat the research laboratory as a 'black box', the internal workings and immediate products of which are hidden from view, but which has calculable effects on industrial activity and generates measurable benefits. Much can be learnt by observing the properties and determining the parameters of such a phenomenological model provided that these can be discerned through the inherent noise in the system. But from the point of view of a theoretical physicist, there appear to be the following serious difficulties in this approach:

Wild uncertainty By definition, research is the pursuit of unknown information, of unknown value. Experience has shown that the value – in terms of expected profit – of a discovery or invention cannot be assessed far in advance of overt success in the market. To allocate resources to research is to enter into a highly speculative undertaking. The prize can, occasionally, be enormous, but the chances of winning are slim.

Under such circumstances, if the 'gambler's ruin' is to be avoided, the only rational policy is to spread the risk as widely as possible over many independent projects. The nearest material analogy is with the search for gold. Conventional large-scale industrial research and development works on the gold-dredge principle – small calculable profits are made regularly by averaging over a vast quantity of gravel containing many small grains of ore. But the whole goldfield was originally discovered by a lone prospector, and its richest lodes exploited haphazardly in the rush which followed. The unreasonable optimism of the solitary gambler and the incompetent extravagances of the mob who try to follow him may be much more significant factors in the initial stages of research and innovation than the rational calculation of self-interest. One sometimes suspects that the scientific instrument makers, like those who got rich by selling whisky and potatoes to the 'Forty-niners', know where the real profits of research are to be gained.

Long-run unpredictability An important innovation usually takes one or two decades to mature. This is too long for successful prediction by extrapolation from existing trends. Even such large-scale average variables as electricity demand or oil prices cannot be predicted reliably over such intervals. Non-linear interaction terms cannot be measured with sufficient accuracy, and imperceptible beginnings – for example, xerography – may have immense consequences.

The analogy here is with meteorology: try as hard as we like, we cannot predict the path of a hurricane ten days in advance. The short-term difference equations used in accountancy and business management would not perceive such singular solutions, or would reject them as inherently unreasonable. Perhaps that is why expenditure on basic research always looks unprofitable to cost accountants. The 'pay-off time' lies in the misty distant future where nothing can be sure, and therefore must be excluded from the analysis.

Many-body effects Modern innovations are born into a multiply connected, strongly interacting social web. The 'non-additivity' of knowledge is paralleled by the non-additivity of related innovations. As in the theory of the liquid state, one cannot isolate the motion of a single atom from those of its immediate neighbours; the approximation of treating the financial, commercial, or technical environment as more or less uniform goes hopelessly wrong. This is evident in any proposed technical solution to the urban transport problem.

Immeasurability of benefits If knowledge and research cannot be evaluated by the producers of wealth, perhaps the benefits can be assessed at the point of consumption. Unfortunately, relative monetary values, that is, preferences, cannot be established over time. I may prefer a current Jaguar to a current Mercedes at the same price; but the car I buy now is not the one my father bought in 1933. Penicillin at a guinea a box is an infinitely better bargain than the pink pills which my grandfather would have prescribed at the same price. Innovation acts on the same products, or to fulfil the same or new needs, through time, over intervals in which no man remains himself entire. In a technically changing economy, there are no invariant standards of taste or performance against which to measure benefit.

The thoughts stimulated by reading this book are evidently rather pessimistic concerning the possibility of constructing a strong and effective economic theory of research and innovation. There seem to be fundamental difficulties in trying to quantify the economic consequences of this immensely powerful social agent – difficulties which cannot be caused to vanish by pretending that they do not exist. This is not to say that economic analysis of research and development is wrong – merely that, like physics applied to biological systems, it answers very few of the vital questions, and is somewhat too coarse-grained to take account of the significant structural features which really determine what goes on. Perhaps the authors have shown the merit of their accomplishment by provoking such fundamental doubts.

26

From Parameters to Portents – and Back*

Quantitative indicators tell us something about science – but not everything.

Introduction: 'disciplined eclecticism'

Robert Merton has advocated 'disciplined eclecticism' as the mental set to be adopted in approaching a topic as complex and inchoate as science indicators. The present paper seeks to exemplify this attitude by suggesting the variety of ingredients that might go into the recipe for such indicators, baked to firmness by persuasive argument.

On the one hand, it is important not to hobble the imagination by attending only to a few conventional statistics that happen to have proved useful in policy making or in academic discourse; on the other hand, the utility of any particular indicator depends ultimately on the accuracy of the observations on which it is based, on the validity of the unstated assumptions by which it is accompanied, and on the logical consistency of the further processes by which it is reduced to operational form. Except to the bigot, eclecticism and mental discipline are not contradictory qualities; but intellectuals are often reluctant to mount attacks on a complicated subject from many sides, in the knowledge that these cannot be coordinated by a simple unifying theory. Whatever they may be or may in due course become, science indicators can only be treated pluralistically, as no more than a miscellany of measures of the secondary characteristics of the multifarious, multidimensional human activity that is loosely referred to as 'science'.

The basic premise of this paper is that the processes by which a social indicator may be produced and interpreted can be broken into a sequence of characteristic steps which are worth separate attention. *Logic* is too strong a word for the relations between the successive stages of arguments

* From *Toward a Metric of Science: The Advent of Science Indicators*, edited by Y. Elkana, J. Lederberg, R. K. Merton, A. Thackray and H. Zuckerman (New York: Wiley, 1978), pp. 261–84.

that seldom claim more than plausibility or rhetorical force. By assigning to each category an alliterative label, we avoid the temptations and ultimate dangers of sharp definition; the schematization is a metaphorical map, not a blueprint. But in trying to construct such a scheme, we learn a little about some aspects of the central problem, and remind ourselves of many potential errors. Formal citations are eschewed; a whole library of reference material is implied in the topics hinted at here.

Parameters

To begin with, we need some parameters; what features of science can be given numerical expression? Most primitively, we can count the elements in distinct categorial sets. *Men* (including, or differentiated from, women, according to context) can be numbered by head in sets characterized by an immense variety of labels, such as academic qualifications, age, employment and institutional membership. In many cases, social and legal principles permit us to count aggregates of men – that is, corporate *Institutions* such as universities – as elements in larger sets. Since instruments, materials, and other *Hardware* play an important part in scientific research, these too can be counted or measured in reasonably precise terms – for example, how many nuclear reactors are available in India for neutron-diffraction experiments or for making plutonium? On the output side, the conventions of the communication system of science make it quite easy to count such software items as *Publications, Citations* and *Patents* – arranged by categories such as authorship, national origin, medium of publication, or subject matter. Methodological questions such as the weight to be assigned to books or review articles need not concern us for the moment.

The facts of financial accountancy and the fictions of market-exchange equivalence are so compelling that it is usual to lay great stress on the single quasi-continuous parameter, *Money*. It is assumed, for example, that the miscellany of hardware can be reduced to this common measure along with buildings, administrative services, postage stamps and paper napkins for back-of-envelope calculations at the lunch table. Such technical niceties as discounting instrumental obsolescence or assessing the exchange value of the ruble against the dollar introduce much less error than other factors we shall encounter. For example, the pressure of technical sophistication can inflate a scientific budget more grossly than any ordinary economic force.

Pertinence

In choosing parameters, we assess their pertinence. An essential requirement is *Consensibility:* there must be reasonable agreement between different observers on the values to be assigned to the variables. Counting heads and dollars is easy enough, but the boundaries of the categories into which these are to be divided are often ill-defined: witness the controversy concerning the proportion of 'Puritans' among seventeenth-century English scientists. For this reason, it is often convenient to use legal categories ('holders of Ph.D. degrees'; 'budget appropriations for the fiscal year') which are sharply defined according to public criteria even though such sets may not be coextensive with intuitively significant entities.

We take for granted adequate testing of the *Reproducibility* of numerical data, but this does not say much about the much deeper property of functional *Invariance.* In the physical sciences, not every quantity that can be measured is of theoretical significance. We are interested, for example, not in the coordinates of the planets in some arbitrary frame of reference but in quantities like the radii and periodic times of the orbits, which do not depend on the standpoint of the observer and which can be related to one another by simple equations. In the course of determining such invariant parameters, we may be forced to measure according to some arbitrary conventions, but all 'frame-dependent' characteristics eventually must be eliminated from the final theory. It is easy to remark, philosophically, that this process of elimination will itself depend on theoretical assumptions; our justification, as always, will be the ultimate consistency of the whole mathematical model with a wide range of apparently unconnected observational data.

In dealing with social and psychological parameters, the lack of well-established procedures for the reduction of data to invariant form is a grave handicap. Statistical techniques such as analysis of variance or factor analysis do not validate the measures they generate. Consider, for example, the attempt to measure the scientific 'ability' of a number of individuals by peer assessment. Out of a diversity of opinions, it may not be difficult to produce a statistically significant rank order; but this does not mean that this parameter ought to appear in a functional relation involving other such observable phenomena as 'quality of research'. In every case, this is a question to be tested on its own merits, and not to be taken for granted. For this reason, the results of Delphi studies, although often extremely interesting in their variety and operating as a stimulus to the imagination, cannot be regarded as satisfactory ingredients of quantitative indicators. At best, such studies provide information about the opinions held by a particular group of people, but tell us nothing reliable concerning the reality about which these opinions are expressed.

A highly desirable characteristic of a parameter is *Precision*. It is important to remember that the standard techniques of mathematical statistics have been devised to deal with quantities like crop yields, which are distributed approximately 'normally' about their modal values. But unfortunately, many of the interesting features of science are very unevenly distributed. Thus the statistical averages, even over large samples, can be entirely misleading. This is known to be the case, for example, with 'scientific productivity' (for example, publications per author), which has an extremely skewed distribution. Because a very small proportion of authors produces a large fraction of the literature, any central statistic, such as the average number of publications per author, becomes quite meaningless. Indeed, if one takes the intuitively quite plausible view that science makes progress mainly through a few exceptional contributions by a few persons of outstanding ability, one must be sceptical about the significance of nearly all statistical parameters which will inevitably be dominated by the pedestrian products of a mass of uninspired research workers doing 'normal science' with little prospect of great success. Until this fundamental structural feature of the knowledge industry is well understood, we must be extremely cautious in the arithmetical reduction of data by conventional methods. As Gerald Holton has emphasized, it is essential to preserve the 'fine structure' of the data, right through to the final interpretive stages of the investigation.

The same warning is called for against the crude assumptions of *Linearity* or *Aggregability* implicit in some parameters. One may naïvely suppose, for example, that the more people at work on a certain problem, the more rapidly it will be solved. But that ceases to be true when the literature of the subject has become too bulky to be properly surveyed and where the administrative demands of large research groups divert the most competent scientists from actual research, or when the need to produce a large number of apparently original contributions displaces – toward a mass of trivial elaborations on an irrelevant theme – the primary goal of solving a genuine problem. Here again, the aggregation of numerical data is no substitute for first-hand knowledge of the social, psychological and cognitive realities they are supposed to represent.

Particularly serious errors arise from the aggregation of data that are not strictly *Commensurable*. This is almost inevitable when everything is reduced to money. For the purposes of financial accountancy, it is perfectly legitimate to lump the stipends of the professors with their annual expenditures on technical equipment and services; but that does not mean that trained intellects are worth precisely what they cost to hire. Within narrow limits – that is to say, within a particular discipline, in a country at a particular level of economic development, over a period of a few years – there may well be a conventional schedule of equipment for a research laboratory that ensures a fairly uniform expenditure on material facilities for each active scientist. In those circumstances, a comparison of total research budgets of institutions or countries is a rough measure of their relative research power. But such a comparison would be nonsensical under wider circumstances; what, for example, would be the proper correction factor for the ratio of the professorial stipend paid by Yale University to J. Willard Gibbs in the 1880s to the salary of a computer programmer at that same university today? Or, would it be fair to say that the research potentialities of the US National Accelerator Laboratory are greater by a factor of 10^4 than those of the physics department of Calcutta University in the great days of C. V. Raman and S. N. Bose?

Knowledge as output

The foregoing objections apply with irresistible force to attempts to measure the output of science. Considered as an 'industry' within society at large, science makes characteristic contributions to technical innovation, economic growth, educational development and other forces or symptoms of historical change. It provides benefits that can be crudely assessed in economic terms. Unfortunately, however, those benefits are often long-term, generalized and non-specific: it is extremely difficult to attribute a particular quantifiable benefit to a limited body of science whose input cost can be estimated, or to assess all the economic benefits that have been gained over a longer period from a particular piece of research. These material benefits flow from the generation by science of an intermediate product – *Knowledge* – which becomes available for common use, often at negligible cost to the user. If knowledge itself could be parameterized, the theoretical basis for cost–benefit indicators for science would be immensely strengthened.

But clearly knowledge is not an economic category that can be quantified. This is not merely a statement of prejudice against attempts to achieve such an end; it rests upon simple fundamental principles. Any scheme for the quantification of knowledge would constitute a theoretical representation of certain processes in the real world. Within such a scheme, the logical relations between symbols standing for real operations like the creation of knowledge and its communication, aggregation and use, must be, of course, essentially equivalent to relations between the real processes they represent. Thus, the abstract algebra of the symbolic system – the logic of the theoretical model we have constructed – must be isomorphous with the intrinsic relational structure of the reality it is supposed to depict. A model that violates this principle can produce nothing but nonsense.

Whatever it may be, the intrinsic logic of knowledge does not constitute an algebra with typical arithmetical properties. Suppose, for example, that we have decided to assign the value x to a particular item of knowledge, A (for example, 'the fact that Mars is covered with craters' is worth 293 *Kans*). What arithmetical manipulations might we apply to this number? Suppose we have assigned the value y to some other item, B; would it then be correct to say that knowing both A and B is worth $x + y$? Obviously, this would depend entirely on the context. If B happened to be sure knowledge that craters on the Moon are caused by the impact of meteors, then the combination is obviously worth far more than if B were some equally valuable but irrelevant fact such as the structure of DNA. The real process of 'adding to knowledge' is not adequately represented by adding the corresponding quantities arithmetically.

Again, what would be the counterpart of the arithmetical process of subtraction? Can a piece of knowledge be taken away, once it is known? The acquisition of knowledge is time-dependent, and essentially irreversible, so that the quantity held by any one person must be a monotone increasing function of time. Or did those physical chemists who had proposed theoretical models to explain the mysterious properties of 'anomalous water' lose a few units of knowledge when it was finally shown that this substance was not a genuine thermodynamic phase – and did the sceptics gain significantly when their persistent doubts were thus confirmed?

Presumably, the process of publishing a new discovery is akin to multiplication; when the first photographs showing the surface topography of Mars appeared in the newspapers, the knowledge store of every physicist in the world must have increased by 293 *Kans*. Is this large quantity of knowledge – say, 293 *MegaKans* – in some sense equivalent to what might be found in a library of 29 300 volumes, each containing about 10 000 *Kans* of knowledge?

The confusions and paradoxes suggested by these elementary examples cannot be resolved by mathematical devices such as giving to the combination of two pieces of knowledge the value of the arithmetical product xy: taking logarithms, we are back at the equivalent of simple addition. It must be emphasized that the logical structure of 'knowledge processes in society' determines the algebraic structure of its symbolic representation; it is scarcely necessary now to invoke standard theorems of mathematical logic to show that this 'knowledge calculus', whatever it may be, could not be reduced to any recognizable form of arithmetic. It is the responsibility of the applied mathematician to represent faithfully what he finds in Nature, not to maim reality to fit his own petty systems of thought.

It is tempting to try to quantify knowledge by referring to its exchange value on the market. The buying and selling of secrets is one of the oldest of professions and it gets honourable mention in economic treatises. But a secret is not a piece of knowledge; it is an item of information, utterly bound to time, place, and opportunity. The price paid for a secret is a measure of its catalytic power in an unstable system. Its knowledge content may be utterly banal ('The King spoke to the Ambassador this evening') and lose all value in a brief time. The calculus of secrets may well be of some interest (thermodynamics of irreversible processes, catalytic reactions far from equilibrium, catastrophe theory, decision trees, nonlinear dynamics?), but it is quite irrelevant to the theory of scientific knowledge.

More soberly, we might try to follow up a remark by Kenneth Boulding: 'It is certainly tempting to think of knowledge as a capital stock of information, knowledge being to information what capital is to income'.[1] Information theory is a well-developed branch of applied mathematics of great practical utility. But the concept of information is there defined in a very strict sense; it is limited to communications whose content is confined, *a priori,* to a finite set of possible alternatives. In this technical sense, and in general usage, 'information' is distinguished fundamentally from 'knowledge' by the former's restriction to a particular context and language of communication. Nobody need doubt that a list of Stock Exchange prices, for example, has a precise information content (that is, in 'bits') and that it is a commodity with an ascertainable exchange value for an habitual purchaser. But its ephemeral life and its coded form show that it is only information; all the Stock-Market ticker tapes that have ever been telegraphed do not add up to our knowledge of that particular institution – which encompasses theories, generalizations, historical facts, anecdotes, mathematical models and many incommensurable quantities, qualities and images.

This is the reason why 'publications' cannot be treated as the quantifiable output of science without considerable reservations. Scientific papers are, of course, very unequal in quality, so that this parameter lacks precision – but that is not the main objection. The fundamental point is that a scientific publication usually communicates information; it is bound by publication date and by subject matter to a definite context, and it is written in the narrow vernacular of a particular scientific discipline. Whether it merely reports experimental data or tentatively sets out new theories, the message is nearly always restricted in advance to one of a set of already formulated alternatives. Genuine novelty in the primary scientific literature is so rare that most papers read to the expert like greetings telegrams; from the title, abstract, and footnotes alone, it is apparent what each one is about – so much so that the American Institute of Physics has been publishing a journal of scientific papers consisting only of title, abstract, and footnote references! It is the accumulation, mutual interaction, and eventual transformation of the mass of information by various intellectual and social processes that turn it into knowledge, which is what we are really seeking.

Patterns

Having determined an appropriate set of parameters, how might we construct an indicator? Into what formulae should the numerical data be substituted? These questions immediately bring to the surface the theoretical models latent in the whole enterprise: significance cannot be assigned to any parameter or indicator without reference to underlying assumptions about science, about society, and about man. Questions of ideological bias, of the protection of powerful interests, of the indefinability of goals, and of the misinterpretation of motive are bound to arise.

What is certain, however, is that none of our theoretical models is sufficiently precise to justify the transformation of the primary data according to complex algebraic formulae to yield simple numerical indicators. In every sociologist and economist there is, one sometimes feels, a physicist trying to struggle free; he reads about Newton's laws of motion, and Maxwell's equations, and Einstein's theory of relativity and wants to put his own data into some devastatingly simple equation like $F = ma$ or $E = mc^2$. In his school textbooks on mechanics, the answers to the problems are numerical quantities like 1427.2 kilograms for the mass of the elephant on the see-saw or 7.3 km/sec for the velocity of the rocket. Surely he must express the answers to sociological or economic questions in similar numbers, such as dollars of benefit per head or Intelligence Quotient points per generation or percentage rates of increase of GNP!

The practical obstacle to this apparently desirable goal has always seemed to be the statistical variance of the data. The precision of the measured parameters is always so low that only mathematical ingenuity and a fair dose of optimism can separate the significant factors from the accompanying 'noise'. Regression analysis, analysis of variance and other statistical devices have often revealed extremely interesting and suggestive correlations between the measured variables, with great utility as empirical indicators and predictors; but the fundamental equations never seem to shine through the murk. It is as if one were trying to discover Newton's laws of motion inside a tank of treacle – or while encaged with a swarm of bees.

Physics enjoyed the luxury of starting with single 'particles' (for example, the planets of Newton's cosmogony). Even in the physics of solids and liquids we deal with ordered, or statistically uniform, aggregates. The real question is whether the 'laws' of sociology are of this kind at all. The search for a 'social physics', or a 'psychophysics', may be entirely vain; what happens in society or inside a person's mind may not be describable in the language of continuous variables linked by partial differential equations. Even if such a description can eventually succeed, we are far from it at the moment and gain nothing by attempting to mould our data into such form. Every arithmetical operation we carry out on the numerical values of our parameters is 'theory laden', in this sense, and becomes a potential source of systematic misconception.

Consider, for example, the number obtained through dividing the total expenditure on equipment in some large institution by the total number of research workers. This quantity – 'equipment per scientist' – looks like a simple indicator of research power, or of technical sophistication, or something. But what does it really mean? In practice, much of the equipment will be shared by a number of scientists, so that each one really has far more facilities at his command than the indicator suggests. However, apparatus may be unevenly distributed within the institution, so that some research workers have very little. But then these might be the theoreticians, who never actually do experiments. Or there may be, in some corner of the building, a genius who is inventing an entirely new device, using very simple equipment – and it may be that the biggest, most expensive piece of apparatus in the institution is actually being used quite fruitlessly by a nincompoop.

The point is not merely that the single ratio contains all the imprecision of its constituent parameters and hides all the detail that would really be of interest; the danger is that by quoting such a number we unconsciously give our assent to some very questionable propositions – for example, 'the more apparatus a scientist has, the better his research', or 'every scientist ought to have his fair share of equipment', or 'if we spent as much per scientific head on cancer research as we do on nuclear physics, we should surely find a cure'. This particular example is not, in fact, a case where any sensible person is likely to be seriously misled, but it illustrates the general principle that even the simplest arithmetical formula, such as a ratio of two parameters, implies a whole theory of the social organization of science.

What should be constructed from the parametric data is not a number but a *pattern* – a cluster of points on a map, a peak on a graph, a correlation of significant elements on a matrix, a qualitative similarity between two histograms, or the connectivity of a topological structure. The term is meant quite generally: it could refer to the qualitative fact of the change of a parameter with time, or even to the observably different rates of change of two different parameters. Two-dimensional patterns are easy to represent on paper and to grasp at a glance, but this limitation is not essential to the definition. The key point is that purely numerical quantities are replaced by geometrical or topological objects or relations, with an overall significance that must be assessed by visual inspection and intuition rather than by mechanical arithmetical criteria.

By the intellectual standards half-consciously adopted by philosophers from the themata of physics, this is a retrograde step – almost as if we were descending to the level of analysis we thought we had left behind in descriptive biology. But this is, in fact, a much more appropriate language than the differential calculus for describing social phenomena, and the patterns we discern in our observations may reflect genuine structural features of the social world. The fine structure of the data is not lost, and abnormal statistical distributions and functional non-linearities are no longer tiresome 'anomalies' that have to be allowed for or 'corrected' mathematically; they show up as features of the pattern, worth serious consideration in their own right. Instead of imposing an alien algebraic structure on the universe of our discourse, we try to draw from it its natural logical topology. This process is never, of course, 'theory free'; but the models we now have in mind are more 'organic' and less mechanical than those implicit in arithmetical formulae. That does not make them softer or less compelling; what may be lost in the indefiniteness of broader qualitative categories is more than made up by the uniqueness of a recognizable pattern or by the many-dimensional correlations implied by the discovery of distinct clusters of objects with many related characteristics. The obvious imprecisions and intuitions in the assessment of the significance of a pattern are less misleading than the uncertainties and systematic errors often hidden in a bare number.

The psychological dimension

What aspects of science demand quasi-quantitative study? What might an indicator indicate? It is all too easy to concentrate attention on problems of immediate urgency where policy decisions are being called for, and to fail to observe phenomena with a longer time-scale. For this reason, it is desirable, now and then, in those brief moments of lucidity between one research project and the next, to take a look around at science as a whole.

In the attempt to give some sort of formal structure to thoughts on this Protean subject, it is convenient to distinguish among the psychological, social and cognitive dimensions in which science has its being: it is an activity of highly conscious persons, corporately engaged in the creation of a body of knowledge. The parameters we measure provide information about *Personal, Political* and *Philosophical Patterns* – at the simplest level, independently, but much more significantly in regard to their interactions. If a science indicator is to have any validity, it must be properly located within this framework of potential qualities and categories.

Personal patterns are those that refer to the individual scientist, as a human being, as a professional expert, and as a member of a group. Attention must be given, for example, to his (or her) *Upbringing*. This takes the investigation back into the basic educational system from the primary school onward, where the foundations of vocational attachment are laid and many intellectual habits are acquired. The relation between general education in science and a specialist education for science is extremely subtle, especially at college level, and strongly affects the important task of encouraging and selecting those few persons with the talents required for successful research. Questions of educational *Opportunity* cannot be ignored, especially where the claims of social equality come into direct conflict with the assessment of intellectual merit and promise. All these issues come to a head in the graduate school, where the aspiring scientist becomes socialized to the forms of his chosen profession.

On the material plane, the conditions of *Employment* in the scientific profession deserve close study. As employees of a large institution such as a university, a government bureau, or an industrial corporation, modern scientists are by no means captains of their fate, and they are seldom in a strong bargaining position for a high salary or security of tenure. The effects of government policy decisions and of general financial conditions in the business world can be serious for a group whose indispensability in the short term is always questionable. The very fact that successful research scientists, by solving the problems put to them, are always in the act of working themselves out of a job contributes a peculiar uncertainty that makes their careers quite unlike those of physicians, lawyers, or administrative officers.

Since research is often combined with *Teaching,* the patterns of academic life demand attention. What does it mean to be a professor? Crude indicators of the burden of teaching, such as student–staff ratios and student contact hours per week, have begun to be heard in the discourse of university administrators, faculty deans and department chairmen. Lectures, seminars, tutorials, practical classes and examinations are not merely educational techniques; they take up a great part of the time and energies of academic scientists. Curriculum reform is as significant for the instructor as for the pupil. Specialization in teaching may be equally harmful for teacher and student, and as much an obstacle to interdisciplinary research as the conventions of the learned societies and 'invisible colleges'. The modern scientist seldom carries out his research alone. Patterns of *Collaboration,* often learnt in graduate school, characteristically differ with time, culture and discipline. Large-scale research, as in space science, calls for qualities of intellect and temperament that might be entirely out of place in a solitary activity such as pure mathematics. Personal attitudes, management structures, the deployment of expertise, training, experimental design and the like must all be matched to the scale of the investigation to make the most of the talents available. Simple facts about human behaviour, known by instruction or learnt from experience by those who manage research laboratories, can easily be lost to sight in the application of mechanical indicators of organizational efficiency. For example, the term *wasteful duplication of research projects* invokes an economic metaphor (conjuring up a vision of, say, a house with two kitchens) and thus fails to recognize the dialectical value of competition and criticism in the pursuit of knowledge.

Among basic scientists, material rewards are supposed to be less important than honorific *Recognition* by the scientific community. Significant questions for the 'health' of science (if that is what we are trying to get indicated) arise from the study of this phenomenon, with its overtones of social anthropology and its undertones of depth psychology. The patterns of recognition (for example, the probability of getting elected to a national academy, or winning a Nobel Prize) change in quality as the scientific community grows in numbers. Traditional rewards of esteem may no longer be adequate – or they may have become so distorted as to do more harm than good.

From recognition springs *Authority,* whether informal – through the 'invisible college' of a discipline – or bureaucratically enforced – as in the hierarchy of a corporate institution. The subtle relation between these two aspects of personal authority in science is by no means well understood. How does intellectual authority come to be recognized? What are its constituents? What does it mean to those who exercise it? What happens when authorities are in conflict? How far should its power extend? These and many similar questions come to mind when one considers the history of science in various countries and its likely future. To give a very simple example, the average age at which scientific authorities relinquish bureaucratic power (whether by compulsory retirement in their sixties, by senility, or by death) may be one of the most significant indicators distinguishing between American and Russian science at present.

These topics do not exhaust the patterns that might be deduced from observation of the parameters in the personal dimension of science. However, the purpose of this brief survey is not to provide a checklist for possible research projects in the sociology of science but to suggest the width and depth of the imagination needed in the search for 'evidences', 'symptoms' and other indicators concerning this most sophisticated of all human activities. As we shall see, the other dimensions of science demand no less eclectic treatment.

The social dimension

What we might call the political dimension of science encompasses far more than the issues that agitate the lobbyists. Modern science is so highly bureaucratized that the pattern of institutions has come to seem more important than the lives of persons. A diversity of *Organizations* – universities, government laboratories, industrial corporations, funding agencies, private foundations, hospitals, technical military corps, publishing houses, learned societies, advisory and testing bureaux, among others – contribute in one way or another to the progress and application of science in society. These are differentiated in size, structure and specialized function, thus providing a rich variety of environments for research and development activity. Once again, overall statistical parameters do not do justice to the range of institutional conditions that actually exist.

The fundamental question about any such institution is the extent of its *Accountability* to other non-scientific political or social powers. Who eventually decides the objectives of research and oversees their attainment? At the highest level of the government apparatus, decisions on science policy may be centralized in a single office, or they may merely be rationalized by compromises among a number of autonomous agencies competing for support. At each downward level we may observe this conflict between centralized and pluralistic tendencies; its outcome has a profound effect on the quality and nature of the science that emerges from the system. Administrative patterns appropriate for one type of goal, such as the development of a new military device, may be quite unsuitable for a purpose like finding a cure for a mysterious disease. Under some circumstances, the best solution is self-government of the institutions by the scientific authorities, using the familiar mechanism of peer review; in other cases strong lay influence, to attain politically chosen ends on behalf of the general public, may be essential.

Institutional accountability may, in practice, be exercised mainly through control of a *Budget* for research. The pattern of such budgets may be the most obvious indicator of actual social priorities – as, for example, in the disproportion between military and civilian research. But the means by which financial resources are made available (whether grants or contracts), the balance permitted between expenditure on personnel and on material facilities, and the degree of detailed control of individual items may all be of the greatest significance. In many countries, for example, cumbersome accountancy procedures aimed against fraudulent misuse of allocated funds may have the effect of greatly reducing the efficiency of research by delaying procurement of small items of equipment.

A significant feature of modern science is the need for enormously expensive experimental *Facilities,* such as rocket systems, nuclear reactors, or research ships. The institutions that grow up to run such facilities, although nominally subservient to the research workers who use them, actually acquire their own autonomy. In prosperous times, with expanding budgets, all is well; but the process of closing down such a facility for economic reasons is not unfamiliar nowadays. Managerial skill is also called for in the organization of cooperation between research teams from independent institutions and in the allocation of appropriate financial resources to joint ventures. Where this cooperation is international, as in

CERN and ESRO, successful research is not the invariable outcome of the attempted collaboration!

The political dimension also includes research by independent corporations for commercial profit. The motivation of such research, its management and its ultimate profitability are widely discussed but by no means well understood. Economies of scale, especially in the development of new industrial processes, are assumed to favour very large corporations; but the spreading of financial risk over many uncertain ventures may be just as important. The great variation of research expenditure between different industries is a matter for careful study; is it mere tradition that restricts fundamental research in food processing to a minute fraction of what is spent in the pharmaceutical industry? The role of government research on industrial processes and the effects of government research contracts within 'private' industry may be highly significant.

Another category of scientific institutions is that of the *Learned Societies,* whose membership cuts right across the boundaries of government and industrial corporations. The extent to which these institutions have become like trade unions, defending the professional status of specialized technical experts, needs to be explored. Does a body like the American Chemical Society represent a coherent group with an authentic voice on sensitive political issues such as the use of scientific weapons in the Vietnam war? How far does a National Academy of Sciences regard itself as an auxiliary of the nation state, as compared with its role as a geographically defined section of the world scientific community? Such questions of fact, or of attitude, are closely related to the 'health' of modern science, at home and abroad.

The learned societies are among the most important institutions for the *Communication* of scientific information. Through their publishing activities and meetings, they largely determine the pattern of the communication system of science, which lies as much in the political as in the cognitive dimension. Journals have to be edited, printed and distributed; conferences have to be planned and managed; books have to be written and published; and large secondary services must be provided for the dissemination and retrieval of relevant information. The learned societies are organs of the scientific community itself, but commercial and governmental organizations are also involved in these activities and must be made responsive to the needs of their customers. These customers are not all scientists; the communication of science to the layman is an extremely important political aspect of science.

Once again, this brief survey of a major dimension of science cannot pretend to completeness. But it may help to show the immense richness and variety of studies that might be undertaken in answer to the broad question 'How is it with science?' Over-emphasis on 'science policy' issues gives too much weight to 'lobby' indicators, which become weapons in power conflicts but are much too crude to reveal the realities of mismanagement, inefficiency, discontent and other pathologies of the science system.

The cognitive dimension

The philosophical patterns of science – its intrinsic cognitive content and structure – are rightly deemed fundamental, even though they cannot be deduced from quantitative parameters. The categories in this dimension are necessarily qualitative; yet that does not render them useless or meaningless. Experienced and knowledgeable observers can come to a satisfactory agreement about a great many facts; most of the natural sciences depend in practice on no greater criteria of 'objectivity' than this level of consensibility.

Although the overall cognitive map of science is quite beyond human capacity to master in detail, it is of extreme interest not only for the mundane purposes of library classification but also as a representation of the natural world. We may scrutinize it, for example, for areas of *Ignorance.* Are there phenomena that remain mysterious, despite intense scientific attack? Are there important questions that could probably be answered if they were given adequate attention? It is often asserted, or assumed, that basic research proceeds with automatic efficiency; the decisions of individually autonomous scientists, familiar with the potentialities for successful research in their own fields, ensure that all good scientific problems are tackled and solved at the appropriate moment by those best qualified to solve them. But this notion of the 'hidden hand' directing the self-governing republic of science makes no allowance for excessive disciplinary specialization or for such sociopsychological mechanisms as intellectual fashions, which often leave large gaps in our understanding. Great unevenness in the overall development of scientific knowledge, with displacement of research goals toward elaborate but trivial puzzles, would be a serious pathological symptom in this dimension.

Small regions of the scientific realm need to be mapped in detail, not only for philosophical or historical purposes but also for the evidence that can be provided concerning the process of intellectual change. The dialectic between the conservative paradigm and the revolutionary innovation is supposedly the mechanism by which truth eventually triumphs, but a good deal of *Error* is also generated and perpetuated in the process. Science's pattern of errors characterizes and delimits its truth in the same way that the pathologies of defect and disease define the healthy body.

Study of the content of individual scientific papers instructs us not only on human fallibility but also on the *Redundancy* and rapid *Obsolescence* of much scientific work. Citation linkages may be used to define the topology of the cognitive pattern in a particular field; they also tell a great deal about the degree of originality that the author ventures to display in his work, the extent to which it is essentially duplicated by other work, and the rate at which each contribution is superseded by new discoveries. This, in turn, is related to the incoherence and *Fragmentation* of research; information is acquired and accumulates in the archives at a greater rate than it can be codified and reduced to knowledge. The assimilation of this knowledge into general culture can scarcely be neglected as the final stage in this process.

The spectrum of *Relevance* of scientific knowledge stretches from basic research to technology. The difficulty is always to weigh potentiality for the future against immediate or past actuality. To use what we already know, in familiar circumstances, is merely the exercise of expertise; research is always speculative with no sure return. Experience and intuition advise on the likelihood of success in solving problems of practical relevance, but tend to emphasize the use of well-tried techniques rather than the unknown powers to be gained by the elucidation of fundamental principles. A major technical enterprise with a long time-frame, such as the harnessing of nuclear energy, demands a very broad mix of research projects, from 'trouble-shooting' in engineering development to the most academic of problems in pure physics. The pattern of this mixture – with short- and long-term research given relative weights, highly speculative and for the moment apparently unprofitable lines cautiously kept open, genuine problems that impede practical advance recognized, and the new discoveries that may quite alter the balance of future technical priorities duly assessed – comprises important cognitive factors in the application of science and technology.

The question of the influence of basic scientific knowledge on technical *Innovation* cannot be answered by crude quantitative studies intended to validate simple mechanistic hypotheses. The depth of the knowledge itself, its analytical or predictive power, the mechanisms by which it percolates into the realm of technique, the nature of education for professional practice, the mobility of individuals between pure and applied fields, the availability of the necessary human and instrumental resources

– all of these and many other factors are involved. Yet this is one of the most important questions for the support and planning of science by corporations and by the state.

To which aspects of human life is scientific knowledge mainly applied? The large place taken by destructive applications – that is, military research and development – should not be neglected. A great deal more could be found out about this subject than is deliberately stated in official documents. Military research ranges from the practical design and testing of battle weapons, through an immense variety of general and technical developments such as aeronautical engineering and electronics, into such realms of higher academic science as the biological principles possibly relevant to chemical and biological warfare. A careful study of the actual patterns of use of such knowledge – for example, in connexion with atomic energy and space research – might be extremely illuminating; the attempt to discriminate between the destructive and constructive potentialities of scientific knowledge is one of the most serious questions of our day. The cognitive dimension of science includes an ethical categorization; the patterns of *Misapplication* are part of the picture, as are the codes of professional ethics against which 'applications and misapplications' are unconsciously measured.

Multidimensional interactions

But science is never so simple as to be represented solely by a *Personal, Political,* or *Philosophical Pattern.* Each of these is, to exploit our geometrical metaphor, merely a projection in one dimension or another of a whole solid object with multiple connexions and strong interactions. The temporary separation of the coordinates can clarify our thoughts, but it must not become a source of error. Indeed, it is precisely the failure to look beyond the cognitive dimension that has blinkered academic philosophy of science, just as the analyses and prescriptions of conventional science-policy studies take inadequate account of personal factors. A similar danger exists that the sociology of science may fail to make sufficient allowance for the cognitive structure of the 'contributions' made by the 'tribes' it studies.

If our indicator patterns are to be realistic, they must be multidimensional and interacting strongly. For example, personal patterns, such as publications, must be related with institutional structures, such as graduate schools, through the cognitive network of citations. To demonstrate interaction among the preceding sections, here let us say that *Misapplications* of science, associated with poor political mechanisms of *Accountability,* may be traced back to faulty *Upbringing* or *Employment* practices. The cognitive structure of science is universal – but to what extent is this universality reflected in international institutions for research, in a world-wide information system, and in personal career mobility across national frontiers? Personal recognition and authority are given for contributions to knowledge, but they are highly influential in the political sphere.

The strength and non-linearity of these interactions are not to be ignored. For example, the political authority granted for an outstanding scientific contribution may become, by the irrationality of human emotions, a barrier against further progress: the erstwhile young radical reigns as a reactionary old tyrant. The applications of science may have proved so valuable that science is planned, subsidized, politically directed and eventually corrupted out of all usefulness. Or reliance on the peer review system for the allocation of financial resources may strongly reinforce the swings of intellectual fashion, to the detriment of progress even in basic science. In choosing our parameters, in the observation of patterns, and in the interpretation of indicators, all such effects must be allowed for.

Perspectives

A numerical indicator or an indicative pattern, standing alone, has little significance. 'What proportion of GNP is spent on R & D this year in the United States? 2.78 per cent, you say. Very interesting, but what does it mean? Is that good, or bad? Should it be more, or less? Has it been increasing or decreasing? What is the corresponding figure in the United Kingdom?' And so on. The data must be given perspective: comparable data must be available for similar systems outside the immediate object of our study.

The natural coordinates of relativity are *Time* and *Space*. As we have seen, the change of a parameter over a period of years may be a much more significant indicative pattern than its measured value here and now. Yet the apparent invariance over several centuries of a characteristic parameter of science, such as the number of pages published in a lifetime by a typically productive scientist, suggests an extraordinary stability of the system, with a basis in very deep and permanent features of human psychology. And when we see change in a parameter that has previously shown great stability – as with the recent flattening of the growth rate of science which had seemed so uniform for several centuries – we know that profound structural changes must be occurring within science and in its relations to society.

But lack of data may make the complete historical perspective difficult to achieve. How deep in time should we seek to delve? This must depend on the phenomena we are studying, which in many cases have their own natural epochs. Little can be gained, for example, from trying to measure the influence of industrial research laboratories before the mid-nineteenth century: such an institution scarcely existed before that date. Again, despite several instructive episodes from earlier times, the mobilization of scientists for war began only in the First World War. Many lessons can be learned from occasional glimpses of the rudimentary forms of such a phenomenon in historical contexts where it could not yet come to maturity; but these lie outside the scope of an indicator.

Yet we must look at present circumstances with the full perspective of time, to see the growth of a parameter or a pattern from its very beginnings. That is why so much of what is done in the name of the sociology of science fails to give a convincing representation of the scientific life; many of the personal and cognitive patterns observed today go back to the seventeenth century without essential alteration, and cannot be explained in the restricted social vocabulary of the contemporary 'organization man'. The information explosion, or crisis, seems a phenomenon of our own times, yet the abstract journal has been a necessary service for at least a century. Expensive instruments like those of Tycho Brahe or the mammoth sundials of the Maharajah Jai Singh in North India have always been characteristic of astronomy, although it would be a mistake to confound the managerial practices of modern space research with the aristocratic authority of a wealthy amateur like the Earl of Rosse. From the history of science, in general, we discover the characteristic time-scale of those features of the science of today in which we happen to be interested.

The geographical coordinate is equally important. It is all too easy to take a parochial point of view and see the development of science in one's own country as unique – in fact, so thoroughly unique as to appear entirely normal. This is a particularly serious matter for the United States, whose scientific output is so large a fraction of published world science that it scarcely seems necessary or possible to compare it with the output of other countries. *Per contra,* smaller countries tend to denigrate their own style of science or to imitate American attitudes, as if 'bigger' were surely better. But the United States is a country like any other, and its science can scarcely be so excellent in all respects as to have nothing to learn from the way things are done elsewhere. The significance of an indicator as a statistical quantity cannot be assessed without some idea of its contemporary variance.

International comparisons in science are made easy by the universality of science. Standards of scientific achievement are widely shared; ideally, the 'invisible colleges' know no frontiers. International journals and conferences set the level, and the best work travels freely in English or in translation. The ancient tradition of the wandering scholar and modern practices of international scientific and technical collaboration give rise to considerable job mobility; the typical European professor has spent several years of his life (not always voluntarily!) in countries other than his own and is quite familiar with general conditions of academic employment, teaching, laboratory facilities, and so on. In spite of the most extraordinary variations in the background culture, the tasks of the scientist or technologist are much the same whether he works in Benares, Baku, Birmingham, or Buenos Aires. For this reason, comparisons of science parameters and patterns among different countries are much more meaningful than comparisons of social or economic indicators.

It is important, nevertheless, to compare like with like. Since the level of *Economic* development is a dominant factor in scientific activity, little can be gained by matching an indicator measured for Indonesia with the corresponding indicator for Sweden. But comparisons between Indonesia and Nigeria, or between Sweden and Australia, might prove extremely instructive. At the same level of national wealth, *Political, Religious,* and other cultural factors come into play; it would be interesting, for example, to make a comparative study of Romania and South Korea – countries of about the same size, with essentially totalitarian governments, industrializing rapidly out of a semi-feudal peasant economy. What are the really significant factors? What are the successes and failures on the way to creating a viable scientific community? At the most advanced level of industrialization, an honest appraisal of the parameters and patterns of science under the would-be socialist policy of the Soviet Union would provide a most valuable background against which to see in perspective the achievements of capitalist democracy in the United States, Western Europe and Japan.

Problems

In using the word *indicator* we imply something more than an interesting theoretical parameter: we are suggesting an evaluation – the answer to some problem. The nature of the indicator and our attitude toward it are not determined objectively, as if in a purely 'scientific' investigation; instead, they depend on our situation, the power we exercise, the position we might wish to defend, and our present values and hopes for the future.

From the point of view of the general public, for example, a science indicator or indicative pattern would naturally be related to the question of *Use.* What's to be got out of science for the man in the street? Almost always this character of sociological fiction is presumed to have purely material needs such as non-stick frying pans – although he too might get a kick out of knowing that Mars has craters like those on the Moon.

The politician or industrial manager who sees it as his job to maximize material economic growth may try to get a yet more refined indicator, relating to *Profit.* In the very short run, and making large allowances for the speculative element, this is a legitimate exercise, capable of concentrating the mind on essentials – although the numbers arrived at by cerebration are seldom confirmed in the event.

In much the same spirit, if we treat science as a source of technical invention and innovation, we might ask about *Efficacy* of research and development. This would be the problem facing the research manager, who must try to deploy his resources of men and money to the best effect to produce an output unquantifiable in strict financial terms but assessable in terms of general value. The familiar issues of 'science policy' in the political sphere revolve about the same concept; Alvin Weinberg's 'criteria for scientific choice' are prescriptions for efficacy indicators in this sense.

For the professional scientist, however, a deeper concern might be for the *Health* of science: potential use may well be threatened by present-day trends within the scientific community or in its relations with the laity. The scientist may feel that long-term potentialities are being sacrificed for short-term gains, and he may look for possible pathologies in less tangible indicative patterns. Comparative measures of the quality of the output may not prove very instructive; this is where the internal sociology of science in all its subtlety should find its application. Unfortunately, there is a catch in this: undue concern about one's health is a symptom of hypochondria, a thoroughly crippling disease!

The personal dimension should not be disregarded; how strong is the *Satisfaction* to be gained from doing research? On the face of it, this seems no more than a subcategory of the health or efficacy indicators, but it is really a dominant factor. Until quite recently, the moving force of science has been individual motivation – the satisfaction of following a vocation rewarded by the social tokens of recognition and esteem. The problem of satisfying these psychological needs is fundamental.

And at the heart of the matter is the question of *Responsibility.* Knowledge is the product of independent minds. Their freedom is essential to the health of science. But that freedom must be conscientious, related to human use. In the patterns of science we need to see indications of this delicate synthesis between these antithetical psychological and social forces and constraints.

Portents

Indicators are thought to be valuable as a guide to action. They inform us about the future. Or rather, they inform us about the past and the present; we extrapolate them, or policies we can derive from them, into the future. In the hard sciences, such an extrapolation is called a prediction; the unconscious model is a prediction of an eclipse – the utmost, the most convincing demonstration of intellectual power, coming straight from the heavens themselves.

In the softer sciences, this model is, alas, no more than a misleading metaphor. Human capabilities for precise numerical prediction are gravely limited. In economics, they have become laughable; in political sociology, no pretence is made. So what could our indicator foretell? At best, what we might call trends or, with a tinge of human emotion (since we really care), portents. Nothing for use, be it understood; but enough, if we are wise, to suggest corrective action.

In a fortunate era, the indicated trend may simply be toward more of a good thing – that is, *Progress*. Except for certain contradictions inherent in our humanity, progress would be highly desirable; but a good thing for one person often turns out to be a bad thing for another. Knowledge, judged to be a universal good of which we cannot have too much, is an ideal parameter by which to register progress. Indeed, short of a Dark Age and a wholesale burning of books, archival knowledge is almost bound to grow; hence, progress is continual and irreversible. More subtle indicators, bearing more instructive portents, are called for.

In economic terms present science may be regarded as an *Investment* for the future: to the eye of the financier, past profitability portends future pay-offs. Broadly speaking, this also is fairly sure, although the cost of producing the knowledge might conceivably exceed the benefits. In practice, the element of uncertainty turns the investment into a gamble; but the prizes keep on growing, the average pay-off over many ventures makes it a good bet.

Or could it be that the portent of our indicators is toward *Saturation?* The number of scientists, the amount of research, the output of knowledge no longer grow exponentially, as if forever. As must happen, they reach a certain level, and stop there. This is no mere nightmare; it seems to be happening, and it suggests many consequences.

Does the system have built-in mechanisms that will stabilize it at an optimum level? Might the trend of the second differential coefficient – that is, the rate of change of growth – itself be steady, so that the curve is passing through a peak and will then turn down? None of our present indicators is sufficiently precise to distinguish between saturation and *Decadence;* there are symptoms of the latter in the intellectual dimension.

Suppose that we had energetically set about determining indicative patterns for the health of science; then we should in due course begin to be concerned about portents of *Disease.* Could these be a source of decadence? More indicators, please – and a better understanding of how science really works.

Eventually, sooner or later – in periods of decades rather than centuries – the past patterns of change, the extrapolated and subsequently verified portents, will have aggregated into something new, a *Transformation* of Science. The academization of research in the German universities in the 1830s and the industrialization, bureaucratization and militarization of science since the 1940s are evidences of past transformations. Like auguries, which could be read only by the high priest himself, the indicators will not reveal this future plainly to untutored eyes, yet should not altogether deceive the scrutiny of the wise.

Feedback

The problems that affect us, and their portents, determine our choice of parameters and the interpretation of the patterns they indicate. The enthusiast for research looks for *Satisfaction;* his indicators will stem from *Publication* parameters, through personal patterns of *Recognition* for dispelling ignorance. But the politician is concerned with the *Efficacy* of research and will measure it in terms of *Money* for *Institutions* and *Facilities.* In a policy conflict each party may look to its chosen indicator for support. Thus the programme of short-term technological development proposed by the politician as an investment is countered by the research man's evidence that it would lead to intellectual decadence. The categorical analysis attempted in this paper is not a simple multidimensional tree; the path from the parameter to its portent closes on itself.

For this reason, the most natural graphic representation of the categorical framework of science indicators is circular – a 'mandala' such as Figure

1. But the subjectivity of the choice of an indicator as a particular lobby's battlecry does not reduce the whole activity to mere rhetoric. Taken as a whole, science remains a distinctive, organic, highly interactive social mechanism with its own level of objective reality. Specialized indicators, or indicative patterns, can inform us correctly about particular aspects of this mechanism, crudely abstracted and simplified, lacking in predictive power or theoretical precision, but instructive none the less. Provided that we keep in mind the whole framework of categories of which these are such limited samples, we need not be deceived. Particular indicators may involve circular arguments, but the whole enterprise is not spherically senseless.

References

1. Standards for the Performance of Our Economic System (Discussion), *American Economic Review* 50, No. 2 (1960).

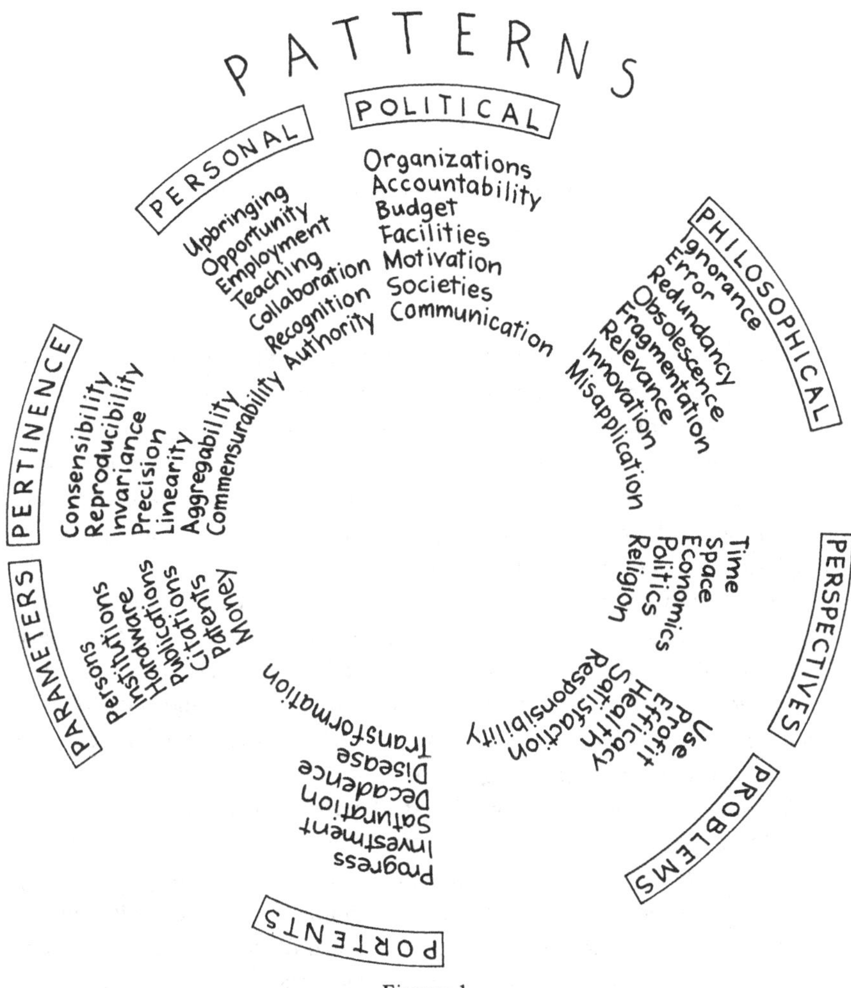
PATTERNS
POLITICAL
Organizations
Accountability
Budget
Facilities
Motivation
Societies
Communication
PHILOSOPHICAL
Ignorance
Error
Redundancy
Obsolescence
Fragmentation
Relevance
Innovation
Misapplication
PERSPECTIVES
Time
Space
Economics
Politics
Religion
PROBLEMS
Use
Profit
Efficacy
Health
Satisfaction
Responsibility
PORTENTS
Progress
Investment
Saturation
Decadence
Disease
Transformation
PARAMETERS
Persons
Institutions
Hardware
Publications
Citations
Patents
Money
PERTINENCE
Consensibility
Reproducibility
Invariance
Precision
Linearity
Aggregability
Commensurability
PERSONAL
Upbringing
Opportunity
Employment
Teaching
Collaboration
Recognition
Authority

Figure 1

27

'Bounded Science'*

What will be the effects of constraining the growth of science?

It seems as if science has always been expanding. Not only has knowledge accumulated; every measure of scientific activity – persons, costs, publications, technological influences – has been growing exponentially for several centuries, with a doubling time of a few decades. There have been doldrums, in one country or another, for various reasons at various times, but the general impression of being in a rapidly growing enterprise has long been characteristic of the scientific life.

This growth has now had to stop. Economics and demography forbid that more than a small percentage of the wealth or population of a society should be allocated to the pursuit of knowledge. This saturation level has been reached in the most advanced industrial countries, and science must accommodate itself to new circumstances where there is no longer the cheerful prospect of continual growth.

Academic science in the United States reached this level about ten years ago. Measured in 'constant dollars', federal expenditures on basic science and university expenditures on research and development have grown very little since 1968: '. . . although support remains at high levels, the atmosphere within which American science is conducted has changed from one of rapid growth and expansion to one of stability and even some contraction. Adjustment from the expansionary conditions of the fifties and early sixties is the basic circumstance that American universities have faced for nearly a decade.' In Europe, this transition from 'expansive' science to 'bounded' science has taken place more recently, but the same generalizations are valid. The S-curve of the logistic function turned over rapidly and the era of saturation foreseen long ago by all realistic observers is definitely upon us.

The question to which the book by Professors Smith and Karlesky is

* Review of *The State of Academic Science: The Universities in the Nation's Research Effort* by Bruce L. R. Smith and Joseph J. Karlesky (New York: Change Magazine Press, 1977), published in *Minerva* 16, 327–39 (1978).

addressed is whether American academic science still retains its health and vigour. 'Academic science' is, of course, a vast and dynamic human activity, diverse beyond arithmetical aggregation. But with the aid of a grant from the National Science Foundation the authors have assembled facts and opinions which give a reliable answer to this question. Indeed Professor Smith and Professor Karlesky have done a really excellent job. It is seldom that one finds such a well-integrated and carefully argued synthesis of facts and opinions covering such a wide area of social activity. The presentation is clear and direct – easy to read, to understand, to agree with, and to quote verbatim. The numerical data have not been manipulated in search of subtle phenomena, but their interpretation is supported by separate studies of different disciplines and 'site visits' to a wide variety of research institutions. The picture is painted in sober, realistic tones, and seems thoroughly authentic.

American academic science is still healthy and vigorous, and it maintains its world pre-eminence. For those who yearn for the golden days when science was the darling of the country there is not much comfort, but good housekeeping and wise leadership are keeping the most serious immediate problems under control. In a period when the whole university system has been under severe stress, the high quality of basic research has survived remarkably well. In my opinion, this is as true of academic science in Europe as in the United States, although differences in the structure of higher education change the relative weights of financial, political and administrative factors in the process of adaptation.

But the results of this thorough investigation arouse one's curiosity beyond the question it was designed to answer. Is this successful adjustment to changed external circumstances engendering a fundamental internal transformation in the whole style of science? Can we observe trends towards a 'bounded' form of science which will be very, very different from the 'expansive' science to which we are accustomed? The authors of this book cautiously avoid any direct discussion of this longer, larger question, but they provide us with much food for further thought.

'Bounded science' must be quite different from 'expansive science' in several significant characteristics. In particular, the constraint of a population in a steady state leads inexorably to the conclusion that the scientific community must become disproportionately middle-aged. The preponderance on the campus of youthful assistants and associates over grey-haired professors, so typical of expansive science, cannot continue.

To show the magnitude of this transformation, let us assume, for simplicity, that an academic scientist becomes established by the age of 30, and does not leave his profession before 60. In a rapidly expanding scientific community, doubling in numbers every ten years, the median age would be around 38. This is consistent with the situation in 1968, when the proportion of 'young' scientists – defined as having received the doctorate not more than seven years earlier – in a sample of members of scientific departments of American universities was about 43 per cent. By 1975, this proportion had already gone down to about 27 per cent – and is estimated will fall as low as 23 per cent by 1980. This is not far from what one would estimate for faculty of science in a steady state; this would have a median age of 45, with only about 25 per cent of its members below 38. In other words, 'bounded science' has room for only about half the number of 'bright young people' who would have been found in the 'expansive science' of the 1950s and 1960s.

This is a very crude generalization. It does not take account of differing rates of deceleration in different disciplines; for example, the proportion of 'young faculty' in physics is estimated to be no more than 15 per cent by 1980, compared with 30 per cent or more in some of the 'behavioural' sciences. But the general conclusion seems inescapable. The model of an academic career is only approximate, since it ignores premature death and other less final exits from the academic world in the middle years. But the mortality rate below 60 is not large, and there is no observable trend or putative factor to contradict the basic assumption that, protected by academic tenure, and with few places more attractive to go to, the average university scientist will stay put. There is no sponge to soak up the 'surplus' of older scientists, thus making more room for the promising young ones. They would certainly not be welcome in the research laboratories of the federal and the state governments, which are subject to the same constraints as the universities and are much less able than they are to provide a wide range of teaching, technical, administrative and consultative opportunities for their older employees.

Nor are they likely to move *en masse* into industry. For all the admirable intentions and earnest efforts to make closer contacts between the universities and industry, 'the coupling problems have worsened in recent years' and there has been a 'relative decline of basic research in industry'. For a British academic scientist, continually agitated with demands that something should be done about links between the universities and industry, it is small consolation to learn that they cannot, after all, manage these matters immeasurably better in the United States. What, then, about those marvellous industrial research laboratories 'like Bell Labs' which British scientists are called upon to emulate? The fact is that there were never many places which tried to be 'like Bell Labs', and most of these have now gone back to much more applied research and development under pressure from accountants of the corporation. Genuinely basic research in industry may have been an epiphenomenon of highly expansive academic science, with which it was deliberately competing for outstanding young scientists and engineers as long-term employees in research, development, production and management. In an era of 'bounded' academic science, this motive has gone; indeed, the incentive now is for universities to go in for 'mission-oriented' research in order to attract industrial funds.

One of the most distressing features of the present state of academic science – not only in the United States – is an acute shortage of jobs for scientists who have recently received the doctorate. This is not just a passing phase. 'Only after 1990 is there likely to be a significant recovery in the number of academic openings available and in the 1980s probably less than one in five Ph.D.s are likely to find academic positions' (quoting from an analysis by the late Allan M. Cartter). But this is not a necessary characteristic of 'bounded science', once the transition from the expansive style has been completed. It is scarcely credible that resources will continue to be provided to train so many gifted persons so very expensively for highly specialized careers which they will be unable to enter. In any case, those same gifted persons will be reluctant to enter careers in which they have so few prospects of entry or advancement.

The numbers of doctorates awarded each year in most branches of science have certainly declined quite considerably in the past decade, but it is doubtful whether they have yet fallen to a level consistent with the needs of a scientific community in the steady state. To estimate this level, we might put forward the hypothesis that the needs of 'bounded science' could just be met if every academic scientist successfully supervised the doctoral training of just one graduate student during the whole 30 years of an academic career. But this assumption of perfect replacement takes no account of 'wastage' – such as the employment of Ph.D.s outside the system of basic science and the loss of motivation for research. If we allow an ample factor of three for a ratio of genuine Ph.D. candidates to established academic scientists and assume that an apprenticeship in research takes five years, we arrive at no more than an average, at any one time, of half a pre-doctoral student per established member of the system of academic science. It is impossible to disentangle the current value of this index from the data given in the present book, but it is surely still a good deal higher than the most generous estimate of the effective demand for scientists in the coming decades. An inescapable characteristic of 'bounded science' is that the proportion of pre-doctoral students, like the proportion of 'young' academic scientists, will be significantly less than in the era of rapid expansion.

Money, unlike human beings, does not come in indivisible units which tend to stay around for years and years. The consequences of putting a financial limit on academic science are not so calculable. But the 'sophistication factor' in science is inexorable. The relative costs of apparatus, supplies, services, technical assistance, material facilities, buildings, administrative overheads, etc., rise much more rapidly than the concurrent indices of monetary inflation or 'economic growth'.

Let us consider the most elementary case, of the independent research worker seeking to replace an instrument which is essential to his research. An automated instrument which costs twice as much – in real money – as its predecessor may well give ten times the amount of data, with ten times the accuracy, in one tenth of the time. This vastly improved 'productivity' thus overwhelmingly justifies the additional cost. None the less, expenditure per research worker has to increase.

In principle, there is excess productivity, which can be shared with other independent investigators in the same department. In practice, this potential gain is not realized because the original 'owner' of the apparatus adjusts his research objectives to make full use of the new instrument in the laboratory, whether it is shared or not – and also expects many new instruments, techniques, supplies or services which did not previously exist and which must now be made available. This is now obvious in the 'big science' of particle accelerators and oceanographic research vessels, but the same effect is to be found in the life sciences where 'fields that have only recently become instrumented are on the threshold of requiring more expensive facilities'.

The increasing size of research teams in all branches of science is another measure of sophistication. The results of team research are often published in the names of all the academic members of the team, whether they are teachers on permanent appointment, post-doctoral fellows, or graduate students. But nobody is deceived by this convention about the actual structure of the team. Strong desires for equality and for participatory democracy in the laboratory seldom compensate for the fact that the initiatives, responsibilities – and rewards – are attributed to a few leading members of the group. In most cases the experiment is really being done by one or two genuinely independent investigators to whom the technical labour of the more junior members of the team is subordinated. This is to say that the most expensive and sophisticated instruments available to mankind – human brains and hands – are being used to get the research results required to settle a scientific issue or to add another paper to a list of publications.

Team research is a manifestation of the 'aggregation effect', which draws research workers into larger and larger organized groups – university departments, research institutes, national laboratories, federal agencies, etc. The 'sophistication factor' puts such a premium on the sharing of facilities that only the aggregation of many genuine investigators and their research teams can bring together all the services needed for scientific work of high quality. Each department or institute must exceed a threshold of size and budget if it is to be effective, but this threshold continually rises as knowledge and technique accumulate.

Team research, institutional aggregation and other indications of increasing sophistication have been so striking in the highly expansive science of the post-war years that we fail to appreciate that these are perennial characteristics of the scientific enterprise. Experimental research is always done at the performance-margin of the best available apparatus, and the need to spend money and other resources to achieve this performance is usually well justified. If the quality of 'bounded science' is to be maintained at the level of 'expansive science', it cannot be allowed to regress to sealing wax, string and dirty old Petri dishes in a crowded attic.

The very idea of science in a perfectly steady state, bounded in personnel and funds, is unrealistic. Scientific and technical progress is a dynamical force, continually transforming its own sources. Every practical or theoretical advance uncovers new challenges, new questions to be answered, new capacities of investigation. The natural response is to take up these challenges on an ever widening frontier; this is the expansive outlook which has prevailed amongst scientists for three centuries. The recognition that this acceleration cannot continue is a severe psychological blow which has not yet been assimilated in the scientific world. It is also indicative of a transformation of the institutional framework of research – not simply to a new quasi-permanent state of equilibrium but into a new direction of change. Driven by the 'sophistication factor', 'bounded science' will have to change institutionally almost as rapidly as the expansive science of the past.

One might, of course, attempt to inhibit this change by freezing the system in its present state. Let us consider, hypothetically, an academic community in which funds for apparatus, and assistants, were attached permanently to certain prescribed university appointments – for example, to a list of professorial 'chairs'. To keep within the limits of personnel and funds, the total number of such chairs would have to be fixed, although there might be considerable flexibility in their assignment among disciplines and departments as the map of knowledge changed, and as new incumbents were eventually appointed. But if the resources in support of each chair were fixed, the quality of research would soon fall, defeated in detail by the 'sophistication factor'. One can also imagine the effects of this policy on an academic community the age-distribution of which had shifted towards the steady state, with a relatively high proportion of older, 'more senior' scientists competing for control of these resources. In the modern American or British academic situation, such a rigid policy sounds fantastic; but it is not entirely unrealistic as a description of what has happened from time to time in the past, in countries such as Italy. From the point of view of a senior scientist in a quasi-independent academic appointment, insistence on a 'right' to existing research funds, assistants, etc., is a natural human response to the threats implicit in 'bounded science'.

The only alternative is to bend to the forces of aggregation and sophistication by concentrating the available resources in a limited number of institutions. In practice, this means sustaining a small élite of 'major' departments or universities at whatever level they need for research of high quality, regardless of the consequences for other places.

The academic world has always had its stratification of esteem, which is easily made manifest as a hierarchy of privilege. The best universities have always had buildings, wealth, equipment and services to match their superiority in students and teachers. A characteristic goal of a policy of rapid academic expansion is to try to 'level up' this hierarchy by giving to less esteemed institutions the resources which would enable them to attract able scientists, ambitious to make use of new, large-scale facilities or to lead their own chosen team of promising junior colleagues. Following this egalitarian policy American universities, outside the major centres, were strongly encouraged to create new 'centres of excellence' in various branches of science. In the 1960s, those controlling the federal support of research were very sensitive to accusations of what was euphemistically called 'geographical imbalance', so that universities in regions which had not previously received a large share of funds for research and development were also deliberately favoured, to the extent that, in recent years, 'there are many ways in which to argue the issue, but the fundamental conclusion is the absence of severe geographical imbalances in the distribution of funds by most accepted measures'. The same programme of expansion was also evident in the creation of the new universities in

Britain at this time. Thus the expansion of science was not uniform and proportionate in all their existing departments; large funds were made available for the rapid growth of many minor institutions, or for the creation of completely new campuses which were to be brought rapidly up to effective size in personnel and facilities.

The last decade has seen a reversal of this levelling process; '. . . tendencies towards stratification (or restratification) among departments are increasingly evident'. But social history is irreversible. The new and expanded departments have kept the undergraduate and faculty numbers they acquired prior to 1970, but have fallen back in financial resources for research, graduate students and equipment. Institutions ranked highly in scholarly esteem have received enough support to meet the demands of sophistication and aggregation, whilst less distinguished departments find that they can no longer live up to their previous high expectations. 'It seems probable that the system is operating to sustain the relative strengths of the most eminent departments.' 'In the past decade, strong departments have generally fared better than weaker ones in the competition for funding. Pre-eminent departments typically do not consider the level of federal research funding to be one of their most acute problems: among the less distinguished departments, decline in federal funding is often cited as a key problem.'

It is reassuring – for world science, as well as for American national pride – that this process has taken place without any obvious decline in the contribution of the American universities to scientific knowledge. There is no reason to doubt the general assessment that '. . . research of high quality continues to emerge from the laboratories of American universities across a variety of fields'. Indeed, it would be extraordinary and tragic if any significant decline of quality in this immense and diverse enterprise were observable in so short a period as a decade. Research communities decay as slowly as they grow, following deep cultural rhythms; they can only be seen to die quickly when they are brutally murdered! 'Bounded science' may be suffering some pangs of hunger, but it is certainly not being starved to death.

But this assessment cannot be backed by any objective measurements. The 'quality' of science is an elusive, self-modifying property the variation of which from epoch to epoch cannot be determined. This incommensurability is compounded by the immense range in the quality of what gets published in any one subject in any given year. Works of transcendent originality or technical virtuosity appear in the same journals as the most ignorant and incompetent rubbish. Looking backwards, we naturally assess the science of any epoch by those few contributions which have really helped our subsequent understanding. But this process of selective retrospection is very misleading, for it does not give adequate acknowledgement to the mass of honest, solid, soundly conceived work in which the great stepping stones are embedded, nor can it compensate for our changing perception of the contribution of the best research of any previous period. Thus, even the vaguest generalization such as: 'Achievements of great consequence in American science are evident in many fields . . .' – with which I would certainly agree – is no better than a well-informed impression in which the criteria of assessment are not explicit.

The trouble is that it is not the small proportion of really good science which costs all the money. From the point of view of the politician or taxpayer, the question might be how much we should pay to get the best science of which our country is capable. There must be enough to be sure that all research proposals which could lead to good results are adequately supported, and must therefore include the cost of following many false trails. The financing of science is a form of gambling, in which a lot must be 'wasted' on losing tickets to be sure of a 'pay-off' on the winners. There is no known economic or sociological law governing the proportion of support which has to be given to relatively poor research in order to increase the probability of unsuspected potentialities for success. In other words, another measure of the quality of science in a given country at a given time might be the amount of hopelessly poor research which is not being supported financially, or otherwise officially encouraged, by publicly responsible authorities.

Although the issue has not been argued publicly, and cannot be decided quantitatively, there is a feeling among many well-informed observers that the great expansion of science in the 1950s and 1960s was very extravagant in this respect. In Great Britain and Western Europe, as well as the United States, a great deal of very mediocre research was supported to an extent that did not directly contribute to the projects which seemed most promising at the time. With plenty of money and jobs around, the critical standards of grant-awarding bodies, peer review panels, deans of faculties and departmental chairmen fell too low. Attempts were made, for example, to create new 'centres of excellence' in disciplines where there was a general shortage of properly trained persons for academic appointment and no outstanding scientists to lead them. Other departments, just to meet their obligations in teaching, found themselves appointing the untried products of very dubious graduate schools, who then brought in surprising amounts of grant money to pay for further expansion. It is easy to refer with pride to the superb achievements of American science of that time; an expert analysis of some of the scientific literature of the same period would show how far the lowest level of grant-aided academic research fell below previously acceptable standards.

The 'bounding' of science has evidently had the effect of cutting into this tail of useless scientific mediocrity. The ranking of academic departments and institutions in the report of the American Council on Education[1] is not an absolute scale of quality, but the fact that 'cuts in federal funding, declines in graduate student numbers, increasing difficulties in obtaining institutional support, and the pressures to do more applied work have touched [doctorate-granting chemistry departments in or below the lowest Roose–Anderson ranking] more than those in higher categories', is surely no tragedy in the forward march of science. Since a lot of bad science actually impedes good investigators in their work, a purge in favour of scientific excellence could even raise the standard of achievement at the highest levels, as well as improving science in general. It could be that we are getting better value for money now than we were ten years ago, when the expansion was halted.

This argument assumes, however, that every scientist of ability is receiving the support which his research deserves, regardless of the standing of the department where he works. In principle, this is achieved by the peer review system of the grant-awarding bodies, who judge every application on its scientific merits. In practice, a member of a poorly esteemed university department seldom gets as strong support for his projects as he probably would if he were in a more famous department. This inequity would be enhanced by the 'aggregation effect'. Support for many small departments of low rank is now insufficient to keep them effective in facilities and services; '. . . the equipment problem . . . is of almost unmanageable dimension for the small engineering schools'.

So far, then, academic science has adjusted to the constraints of the past decade by restratification. This adjustment has been remarkably successful in not doing any obvious damage to the scientific output of the academic system as a whole. Looking around today, one might agree that 'vigorous competition among a slightly thinned number of science departments could . . . provide a stable system'.

Restratification has not come to a stop. The 'sophistication factor' is inexorable. 'There are a variety of pressures pushing in the direction of increased stratification; these are related to the internal logic of a field, the practice of funding agencies to focus research energies on a specific target, the scale and cost of modern instrumentation and facilities, or other causes.' Science which is 'bounded' in personnel and funds must either follow this path, or else fall into decline.

How far can this process continue without harmful consequences? This is anybody's guess, but certain conclusions are obvious:

> If the nation were to move too far in the direction of fewer and fewer outstanding university research departments . . . national breadth and diversity [of scientific competence] would be endangered. It is possible that the competitive elements in the system could also be undermined if there were too few universities fully equipped for preeminent research. The nation might then be forced to rely on a narrow base of dominant centers. The competitive pressures in the system can, however, operate quite effectively with, say, 50 strong departments in a certain field in contrast to twice that number. But could such competitive pressures work equally well if there were only 15 serious producers of quality research in a given field? . . . Is it enough for the nation to have a relatively small number of clearly élite centers of scientific excellence?

Most experienced academics would answer, without hesitation, that such an outcome would threaten the quality of science itself. The optimal degree of institutional competition cannot be calculated, but the extreme which approximated to monopoly would undoubtedly be disastrous. If we are to understand the full implications of 'bounded science', this danger must be taken seriously.

From a naïve point of view, science might be supposed to benefit by the concentration of the most powerful apparatus in the hands of the most able investigators. Politicians and businessmen often decry the apparently 'wasteful duplication of research'; doctors, teachers and engineers complain of 'the proliferation of scientific journals'; journalists ridicule controversies about priorities and recognition amongst scientists. In a rational world, so it is assumed, science would be managed to avoid all that untidiness and waste.

The truth is, of course, that this 'wastefulness' is the price which must be paid for the continued reliability of scientific knowledge. This depends not so much on the perfection of apparatus and technique as on the competitive individualism of independent investigators. This independence is a paramount necessity. A university is a proud institution, the members of which are jealous for its scholarly reputation, and exert themselves mightily to protect each other from external pressures. Academic tenure in an autonomous corporate body is the traditional foundation on which the scientist takes his stand, whether as a radical innovator or a reactionary critic. This is the social mechanism by which the scientific attitudes of intellectual integrity, free speech and mutual tolerance are maintained in practice.

It is not clear how these attitudes would be maintained in larger, more centralized research corporations, without the challenge of competent autonomous equals in each field of research. There is no certainty that they would persist, generation after generation, if they were not impressed upon every scientist by the social structure of the academic world. Historians of science, ancient and modern, can point to cases where the academic system of a country has come under the sway of a single élite institution, the subsequent decay of which has only been held in check by competition with foreign institutions of learning. The immense intellectual vitality of present-day American science owes a great deal to the absence of a very small élite of institutions or individuals. Every university department, however distinguished, knows that it must take seriously the competition of its peers. Within the conventions of a democratic society, there does not seem to be any other way of maintaining high standards in research.

In some fields, such as space research and high energy physics, the 'sophistication factor' is already forcing the aggregation of facilities in an alarming way. There are certain experiments, for example, which can be done at only one centre in the United States, with the permission of a single committee managing the research programme of a space-probe or accelerator. The committee will undoubtedly deal with such a proposal with the utmost scrupulousness; it must nevertheless be realized that a truly independent challenge to the results reported from such research could now come only from a research group in another country with access to a comparable instrument. For the present, and the immediate future, the general ethos of competitive individualism, the wisdom of those who manage these great instrumental facilities, and some subtle institutional arrangements built into the charters of the national laboratories, are quite adequate to maintain scientific standards and protect the critical freedom of individual scientists working in these fields. But in the long term, the processes of stratification which are at work in 'bounded science' must not be allowed to run on, unchecked, to a monopolistic extreme.

Financially bounded 'big science' thus faces a difficult choice. Should it strive for the most powerful instrument which is technically possible, knowing that this can only be made available to a favoured research group, or, at best, to a consortium in which the independence of research groups must inevitably be compromised? Or should competing groups each get a more modest but 'fairer' share of the available resources, so as to preserve the intellectual health of the whole scientific community? It is possible that the age of expansive science permitted the support of extravagantly designed instrumentation – of many white elephants and pink palaces of more than oriental splendour – which was much too grand for the job it had to do, or which failed in performance. Until recently, however, the size of any single research instrument has been restricted mainly by technical feasibility; a determined and talented enthusiast could usually obtain the funds he needed to make his dream come true. In the new era of 'bounded science', not everything which could be done, however scientifically attractive, will be able to be done.

The recognition of this ultimate limitation does not however justify a policy of harsh stringency in the provision of funds for apparatus. 'While proportions of NSF grant support spent on personnel, expendable equipment, travel and publication costs have remained relatively stable since 1967, the proportion of grants spent on permanent equipment has declined from 11.2 per cent in 1966 to 7.3 per cent in 1975.' This trend has very human explanations; when funds become more limited, officials are tempted to allocate money to pay the costs of essential personnel and to postpone the acquisition of new equipment. Investigators, too, have been reluctantly willing to postpone replacement or to neglect adequate maintenance. However understandable the short-run adaptation to austerity, the long-range consequences of a base of deteriorating equipment for American science are serious. With a research life of three to six years, middle-range instruments are wearing out and becoming obsolete. The 'sophistication factor' cannot be disregarded to that extent without significant effects on the quality of research.

There is no way of calculating the proportion of the total cost of research which 'ought' to be spent on permanent apparatus. Science is not an industry where the future benefits of investment in automated machinery can be estimated in terms of profit and loss. Experience and wise judgement will continue to be the only guides. None the less, some branches of research are quite close to pricing themselves out of the market, even on the world scale, while others are showing monopolistic tendencies which could be equally damaging. To escape the fate of the dinosaurs, 'bounded science' must continually nourish all which is new and vigorous in the 'littler sciences', the intellectual merits of which are often underestimated by comparison with their more spectacular competitors.

The aggregation effect tends to put research facilities into the hands of older, more experienced scientists, and of leading research teams. Since the proportion of older scientists is much higher in 'bounded science' than in the era of expansion, one might expect it to become much more difficult for young investigators to obtain grants for research. However, 'the evidence on this issue is mixed, and no simple generalization accurately describes the situation for young researchers in all disciplines'. This may not be simply a consequence of peculiar wisdom, amounting to a strong prejudice in favour of youth, on the part of grant-awarding bodies. Subtle factors work on older academics on permanent appointment as science becomes less expansive and more of an establishment. For example, graduate students are more than cheap, willing and intelligent hands; they are the scientific progeny of their supervisor, who exhibits paternal pride in their number and their subsequent successes. Without graduate students, the senior academic scientist loses both technical help and a strong motive to continue his research. There is not so much satisfaction, after all, in renouncing 'the small scale, intimate one-professor-and-a-couple-of-graduate-students' style of research, in favour of 'one with a "big science" emphasis, using more technicians and post-doctorates'. He may not be so unhappy as he thinks he ought to be when he drops out of competition for grants.

The intellectual vitality of an academic department depends very much on the scientific enthusiasm of the younger faculty members. This enthusiasm may be deadened by long years of waiting for promotion to tenure. In the era of 'bounded science', titular professional preferment is inevitably slowed down. Provided that younger staff members are adequately paid, and are given enough freedom and security of appointment to pursue their own lines of research, this is not necessarily a real handicap.

The scientific life thrives on the interaction of new persons with new ideas. One of the striking characteristics of the American academic world in the 1960s was the extreme mobility of individuals between institutions. For the most energetic and able, it was easy to gain rapid promotion by moving to an expanding institution in another part of the country. Ideas and techniques were thus 'carried around inside people', whilst many individuals were stimulated to start on new lines of research by a change of their academic environment. If one of the effects of putting bounds on the academic profession has been to reduce the mobility of individuals to quite low levels – the quantitative evidence on this is not known to me – then much will have been lost. It is easy to see, of course, that 'bounded science' in a steady state need not necessarily be immobile in the locations of its individual members. The British and European traditions of seeking external candidates for a fixed complement of vacant chairs produces quite a lot of movement among the most able and ambitious scholars. There is, however, always a strong temptation, under present-day conditions, for the individual to wait for permanent tenure in the same institution, to wait for his elders to die or retire; at the same time, academic appointments committees try to improve the age-composition of their departments by appointing the youngest promising applicant.

It is thus very difficult to estimate the effects of all these factors on the distribution of research effort in the academic world. Numerical indices are almost meaningless, because of the great range of talent among scientists, young and old, and the variation in the ages at which these talents mature, or eventually die, in each individual. It is even more difficult to demonstrate empirically the 'self-evident' proposition that putting bounds on science will make it less adventurous. Even in superior institutions '. . . perhaps the most serious potential danger is the tendency to "play it safe" when funds are tight . . . among the manifestations of [this] tendency is the increasing reluctance of some graduate students to take risks; they prefer a safe thesis topic that is more likely to lead to a certain job'. Is this really true? There is no doubt that 'young investigators must work in a far less encouraging research environment' and 'it is unquestionably much more difficult for the young to establish themselves now than it was ten years ago'. But it is extraordinarily difficult, looking back from the heights of middle age at one's own years as a student, to make a comparison with the attitudes of students of the present day. In a couple of decades, the style and content of every science suffers such changes that there remain no reliable benchmarks for the assessment of the relative degree of 'safety' or 'risk' in research projects. Other social factors, such as general affluence and permissiveness, may be at work, dulling ambition or decreasing the compulsion to live laborious nights and days. One might just as well have confidence in the continued power of science to hold the imagination of a certain proportion of talented young persons in every generation, and to draw them into nearly impossible feats of intellectual virtuosity.

Despite the enormous psychological significance of the ending of the era of expansion, the state of academic science is not uniquely determined by such crude factors as the level of financial support for research or the rate at which the scale of this support is changing. In any given country, at any given time, there are equally important effects of scholarly tradition, institutional structure, political, economic and cultural contexts and other factors. In its post-war prosperity, the morale of American science became almost intolerably euphoric. That mood alone, by psychological inertia, should have carried it safely enough through the last, more sober decade.

More significantly, the 'bounding' of science has not been accompanied by any profound change in the system by which it was supported. There has been no fundamental reform of the 'indefensible', administratively incoherent, politically unaccountable and financially wasteful procedure by which a multiplicity of independent agencies fed grants into the ever-open, thrusting mouths of a large number of jealously competitive university departments. It may have been untidy and costly, but it really worked. And for all its occasional extravagances, American science has kept itself astonishingly clean from personal or institutional corruption. '[Programs] . . . designed to aid special constituencies tend to introduce political considerations into the process of awarding research grants. These are at odds with the principle of competition on the basis of scientific merit – a principle that has served the country well.' Scientific excellence assessed by peer review remains the fundamental criterion for support.

Nevertheless, 'bounded science' is all too vulnerable to the traditional distresses of the government pensioner – political pressure, financial fickleness and bureaucratic pettifogging. It can no longer free itself from external constraints by expanding irresistibly in all directions. As a result of quite rational administrative changes in the federal government, 'the degree of pluralism in research support has declined and investigators have fewer potential sponsors for their fields of inquiry'. Again, for a variety of reasons, 'the grant acquisition process . . . has not merely become more competitive – an unavoidable consequence of a larger number of scientists seeking limited funds – it has also become more elaborate, time consuming and bureaucratized'. And one can now see very clearly a fundamental weakness of the system in not providing the universities with a regular basis of 'umbrella grants', not tied to specific projects, which would make it much easier for hard-pressed institutions to redeploy personnel and resources flexibly and smoothly. These are, however, issues peculiar to the American academic and governmental scene. The scientific establishment is big enough and strong enough now to protect itself from major political follies, and it understands well enough the valuable characteristics of the pluralistic system by which it came into being.

The fundamental issue which arises in the 'bounding' of science is not about 'science and government' – for example, 'How much science should there be, how should it be funded, and who should direct it?' In all advanced countries now, the real issue is the relationship between research and education.

The effects on internal stratification of putting 'bounds' on science are so strong that it is hard to believe in the possibility of maintaining 'a pattern similar to the present system, with graduate education and research of high quality, widely scattered among many universities across the country'. Nor would one wish for 'a continuation of the present pattern of numerous doctoral programs, but all of them reduced in quality, and the general standards of the educational system in decline'. Almost inevitably, the outcome must be 'a pattern of substantially diminished doctoral-level work in many institutions . . . with a relatively few universities engaged in extensive doctoral-level training and associated research activities'.

Academic science expanded with the entire academic system. In most countries, the great post-war increase in the proportion of young persons going on to higher education was not accommodated 'on the cheap' by simply expanding the intellectually poorer academic institutions such as state colleges or technical colleges. Real universities were encouraged to grow, and many new universities were founded. As has already been remarked, the research achievements of some of these new or inflated institutions were not very high, and science itself would lose little if these activities were curtailed. But there is all the difference in the world between an institution of higher education in which the teachers are employed solely to teach, and one in which the teachers are also involved in non-didactic work such as research. In the former case, educational exercises become ends in themselves: the teachers have no framework of reference beyond the disciplines which they once assimilated and no means of connecting themselves or their pupils to other aspects of reality – social, intellectual, practical or aesthetic. It is a crime or a fraud against a young person to put him or her into a collegiate form of education which is no more than an extended version of the secondary school or a hash of half-baked, second-hand accounts of the latest in scholarly wisdom.

Higher education in the natural sciences has always had the virtue that it could always be validated by the touchstone of current scientific knowledge. Even a mediocre teacher automatically internalizes this standard of intellectual justification if he is himself engaged actively in research in his subject. He feels himself to be personally involved in the pursuit of knowledge, and communicates this involvement to his pupils. In other words, science is taught as a living, dynamical process, not as a body of established formulae.

'Bounded science', instrumentally sophisticated and professionally aggregated in superior institutions, can offer no opportunities for continued research for a high proportion of those who must teach science at the level of an advanced secondary school. The actual degree of aggregation and concentration of such facilities will depend on a variety of factors, such as the acceptable stratification of esteem in the higher educational system, the extent to which superior institutions divest themselves of undergraduate instruction and become exclusively graduate schools, the balance between teaching and research in the career of an active scientist, the length of graduate training and research experience required of teachers in the tertiary level and many other administrative formulae or cultural traditions. Specific issues of this kind within the American context are discussed with characteristic shrewdness and sobriety in the final chapter of this book, at a length and in a depth which defy summary. But the authors do not show any way out of the general conclusion that the harmonious symbiosis of higher education and research cannot survive in its present form.

I must admit that I am surprised and depressed by the inevitability of the conclusion thus reached in writing this review. The argument, as we have seen, does not depend on the circumstances of a particular political or social system; American mass education, although showing many of the symptoms of a coming crisis, has admirable powers of adaptation which may preserve its valuable qualities for a long time. The crisis is inherent in the nature of science itself, continually transforming society about it, but also necessarily self-transforming from generation to generation. 'Freedom,' Engels wrote, paraphrasing Hegel, 'is the recognition of necessity.' Within the calculable consequences of putting bounds on science, we still enjoy quite enough freedom for almost all manifestations of wisdom and folly.

References

1. Roose, Kenneth D. and Anderson, Charles, *A Rating of Graduate Programs* (Washington, D.C.: American Council on Education, 1970).

PART FIVE

SCIENCE IN THE THIRD WORLD

28

Some Problems of the Growth and Spread of Science into Developing Countries*

The problems are not entirely technological or economic: they are historical, political, cultural and psychological.

The story of Ernest Rutherford, the poor farmer's son, from the back-blocks of distant little New Zealand, who won a scholarship to Cambridge and became world famous, is still one of the inspirational epics of the life of the mind. Many another young man, from many another far-off land, has taken it as his model, and gone to seek his fortune thus, in the great intellectual metropolises of Europe and North America. The urge to contend and make one's mark in the great game of 'high science' is as strong as ever among the ideals and ambitions of gifted youth.

Yet times have changed; from the point of view of the patriotic New Zealander, Rutherford's permanent migration to England might now be seen as a serious case of 'brain drainage'. In speaking in his memory nearly a century after his birth, it seems appropriate to discuss the converse phenomenon – the growth and spread of basic science, from its original nuclei in the industrial countries of Western Europe, into all corners of the Earth. We hold it almost as self-evident that currents of knowledge, skill, attitudes and techniques should diffuse the culture of scientific research throughout the world, so that eventually the conditions for the pursuit of mysterious Truth may be provided anywhere, from Timbuctoo to Tahiti, from Kamchatka to Kathmandu.

Now, of course, for a developing country, struggling desperately to provide a better life for millions of ordinary men, women and children, the problem of giving an appropriate priority to this essentially exotic activity is not trivial. I am not speaking now of the techniques that pure science has uncovered and created – improved methods of agriculture, the

* Rutherford Memorial Lecture of the Royal Society, delivered on December 2, 1968 at the University of Delhi, published in *Proc. Roy. Soc.* A, 311, 349–69 (1969).

control of birth and disease, rapid communication and transport, and so on – but of research without any more conscious aim than the understanding of how things are, or were, or might be. When money and men are scarce, it is by no means self-evident that an institute of theoretical physics, or molecular biology, or archaeology, should take precedence over a tractor factory, a hospital, or a school of civil engineering. Even in Britain, which is already as rich in material comforts as any sane man would wish, we argue about these priorities; in a country such as Indonesia, or Brazil, or the Congo, a very strong case indeed must be made for the diversion of any financial and human resources at all into such unproductive channels.[1]

As a first approximation this argument is compelling. Those of us, in advanced countries, who have the good fortune to be allowed to make our living solving the splendid puzzles of Natural Philosophy may be well aware that what we are doing is not likely to be very useful; but we can comfort ourselves with the thought that we live in a society where most of the basic human needs are adequately satisfied, and that we are no more parasitic than such admirable citizens as estate agents, jockeys, bar tenders or advertising executives. But this irresponsible attitude is not permissible in less fortunate societies; in a country such as India or Pakistan, I can see no justification for, shall we say, the neglect of practical animal husbandry for the sake of research in pure physiology, or for the study of the theory of superconductivity in the absence of good schools of electrical engineering. I want to say this at the beginning in case I seem to be yet another pure scientist who has his nose so deep in his silly little field of research that he cannot look out at the tragic real world around him. Indeed, I suspect that the solution of technological problems, which come already loaded with human cares and consequences, is more satisfying than finding the answers to those abstract impersonal puzzles which we dream up for ourselves in our ivory towers; and I simply cannot understand the intellectual snobbery of those silly people who give more credits for the discovery of another meson than for the design of a suspension bridge, and who cannot see that a zip fastener is a far more beautiful idea than a zeta function. Who knows, Rutherford might easily have earned his immortality for developing radio and radar rather than for splitting the atom.

But the utilitarian argument against basic research can be carried too far. In an interesting series of controversial articles Professor Harry Johnson has attempted to show quantitatively that pure science is essentially an extravagance whose applicable results may be bought from abroad far more cheaply than they can be produced at home.[2] This follows from the rationale of free trade and the international division of labour. The developing country benefits more by the export of a few clever people, and by the subsequent import of technical know-how, than it would by attempting to expand its own very inefficient and unproductive local research industry.

But, as others have countered,[3,4] mere economic accountancy is not compelling in such a complex of subtle circumstances. A certain amount of fundamental research *must* be sponsored in a developing country, for a number of excellent reasons which have been clearly stated by Moravcsik.[5]

In the first place the education of technical experts – engineers, doctors, agricultural advisers, even government administrators – cannot be left entirely in the hands of technologists of their own practical kind. Modern engineering, for example, requires the exercise of skills, and the application of knowledge, acquired from pure physics, chemistry and mathematics. The rapid development of new techniques can only be exploited if the practitioners are adequately trained in these fundamental disciplines, by teachers who are themselves in close contact with the latest theoretical principles. If we have learnt anything in the past century about scientific education, it is that only those who are actively engaged in research can truly absorb and retransmit these new principles as they arise. The research work of university teachers is not an extravagant irrelevance to their professional task; it is absolutely essential to their efficiency in passing on useful, applicable knowledge to their pupils.

From a purely practical point of view also, it is impossible to import technical know-how, and to apply it successfully, if one does not have available, locally, a corps of learned men to whom one can appeal for guidance on matters of pure scientific principle. Suppose, as an imaginary example, that we have set up a factory for the production of simple radio sets. The beautiful machinery for making the transistors, imported from America or Japan, does not work properly, perhaps because there is too much copper in the water supply. The engineers study the problem and eventually unearth a paper in the *British Journal of Applied Physics* referring to 'the diffusion of transition metal ions along dislocations in covalent semiconductors'. Their training is adequate to suggest the relevance of this paper to their practical problem – but only an 'academic' physicist, with a profound knowledge of the science of materials, could really tell them just what it meant, how much it could be relied on and whether it contained the answer to their questions. All our experience tells us that such queries cannot be dealt with by correspondence at a distance. As

Crawford has pointed out,[6] local experts can see more clearly the needs and wants of the country. Science-based technology is not a lusty crop of rules of thumb, perfected by evolution over long periods; it is a delicate plant, which thrives only when tended by mixed teams of experts, including those impractical specialists to whom the buck can be passed when fundamental principles are at stake.

Quite apart from utilitarian arguments, there are spiritual considerations. The pursuit of scientific knowledge for its own sake has become one of the major artistic enterprises of humanity. Even though we would deplore the supersession of all other forms of art, sport or religion by such a specialized activity, we must surely recognize that the opportunity to undertake scientific research is one of the important attributes of the Good Society. I cannot imagine any goal towards which a human society might claim that it was 'developing' that did not nowadays give some room for at least a few of its members to participate in this great, world-wide movement. The attempt to understand the genetic code may promise no immediate practical dividends, and may perhaps give direct pleasure to only a small proportion of the population. Nevertheless, like a ballet, a poem, a park or a stained glass window, it embellishes the society that has commissioned it, both for its own sake as a thing of beauty and as a symbol of the time to come when bread will not be our immediate lack. Failure to give at least token support to pure science would be as philistine as banning the public performance of orchestral music, or ploughing up Regent's Park to grow potatoes.

There is, therefore, a very good case indeed for modest encouragement and financial support for basic research in any country on the threshold of industrial development. On the other hand, these arguments should not be carried too far. They do not, for example, justify leaving the choice of basic sciences entirely to the accident of the availability of trained men, and the directions into which their enthusiasms happen to lead them. There are occasions when deliberate decisions of policy can be taken at a relatively general level. Shall we foster ecology or physiology? Should we strengthen plasma physics or materials science? Do we need an institute of molecular biology, or a department of radio astronomy? Surely one should then take account of the relevance of the subject to the economic, geographical, meteorological or social background of one's country. Even within pure science there are branches that are more natural, so to speak, to one region of the world than another, and which are therefore more likely to thrive in the appropriate context. In Alaska, for example, a physicist might find glaciology congenial – while the study of thunderstorms could be left to the fortunate inhabitants of the tropics; in Western Australia, with its vast new mining industry, he ought to become interested in mineralogy – while the Japanese steel industry might encourage him to study the mechanical properties of metals – and so on. This chameleon behaviour is not a matter for formal legislation, but quite apart from its technical consequences it is the sort of shrewd and practical policy that can counteract the brain drain. It gives the local scientist a better reason for staying where he is than mere patriotism. If, like one of the first real scientists I ever knew, you specialize in the physiology of the mammary gland, and if cow's milk is the life blood of New Zealand, then where better to live and work than beside the Waikato River! On the other hand, what could be more demoralizing than attempting to construct theories of superconductivity in a laboratory without low temperature facilities, or more frustrating than trying to look through a large astronomical telescope in the climate of the British Isles!

It would seem sensible, also, to put the main emphasis on what I should describe as 'potentially applicable' science. Of course we do not know in advance what may come of any particular scientific discovery. It is a trite argument, often used to justify the most esoteric activities, to point out that nobody could have known the practical uses of radioactivity, or electromagnetism, where they were first discovered '. . . and therefore, gentlemen, you must of course give me another million dollars in case I discover something equally useful up there in the stratosphere' etc., etc. I am afraid I do not see the necessity. Let us not stand in the way of utterly pure mathematics, astronomy, palaeontology, and other apparently useless disciplines; but don't let us fall into the trap of equating uselessness with snob value. There are numerous excellent, difficult and rewarding branches of pure science which at least attempt some understanding of humanity and the everyday human environment and which therefore promise, some day, to be of real use. Although I have pretended to laugh at my own little game of solid state physics, I could easily trace the connexions between the basic problems we study and such practical activities as the design of radio sets or of bridges.

Indeed, the distinction that is here being made between 'applied' and 'basic', 'fundamental' or 'pure' science is not at all precise, and the attempt to make it sharp and clear is bound to fail.[7] I am using a rough-and-ready conventional classification, which I should not want to defend in detail. The sooner we all face up to the fact that theory and practice are indissoluble, and that there is no contradiction between the qualities of usefulness and beauty, the better. If there is one feature of American science that we should all imitate – I speak for Britain, Europe, and most of the rest of the world – it is their willingness to have both fundamental and applied research going on together under the same roof. Their universities do not despise technological development work, while their industrial laboratories are well equipped with basic research groups. In a developing country, where short-term problems have much the higher priority, the conventional practice of keeping such work out of the universities is quite indefensible; as I have already said, the main function of long-term research without specific goals is to provide the best intellectual and educational background for well-informed technical innovation.

From these relatively sober considerations, it follows that any developing country, once it has emerged from the most primitive poverty, should foster a small amount of basic science. Of course this can be overdone; a research reactor can become as extravagant a 'status symbol' for a small country as its own international airline. But I do not wish to dwell upon the difficult problem of relative priorities and the criteria for choice between technology, applied science and fundamental research. For the remainder of this lecture I shall try to express some opinions concerning the establishment and maintenance of basic scientific work under these difficult circumstances. Although the obstacles are all too evident for those actually involved, very little attention seems to have been given to this problem in all the mass of writing and talking on industrial and technical development. In particular, I want to emphasize the intellectual and psychological aspects, which are neglected in the struggle for material resources.

At the outset, one must have a clear idea of the nature of science itself. It is all too easy to adopt an uncritical, imitative attitude. 'Science is what they do at the Cavendish Laboratory, so we must try to build a laboratory that is just the same.' But any attempt to make a carbon copy of a social institution, under another sky, is bound to end in disappointment. If we are to foster new institutions that really work – productive and creative Indian science, or Pakistani science, or Nigerian science, or Patagonian science – then we must have a proper understanding of the long-term goals and short-term functioning of the parent institution on which they are moulded. To give a ridiculous example, putting on a white coat and peering knowingly down a microscope no more makes a biologist than chanting metrical psalms makes a Presbyterian. It is the inner spirit that counts, not the ritual.

For this purpose, it is essential to begin the analysis at the most general level; indeed, one must first try to define science itself. As I have argued at length, in a recently published book,[8] the most useful definition comprehends the *social* dimension of science. The role of the pure scientist, however eccentric, imaginative, speculative or opinionated, is to make a contribution to 'public knowledge', to help construct a rational consensus. Many familiar dichotomies – experiment and theory, creation and criticism, accidental observation and deliberate search, logic and intuition, master and pupil, freedom and authority – can live together without contradiction in this definition. And, as I hope to show in this lecture, one can deduce from it some quite powerful guiding principles for the practical management of scholarly affairs.

This point of view implies, conversely, that there is no very special 'philosophy of science' that must already be latent in the culture of a developing country before a scientific community can be established. It is sometimes argued that the philosophical, religious or cultural climate in certain regions of the world is so hostile to the scientific attitude that research could never thrive there without a complete psychological reorientation.[9,10] 'In Jub-Jub Land,' this argument runs, 'the study of botany would be quite impossible, for everybody is brought up to believe that oak trees spend the hours of darkness flying to the Moon', or 'How could you teach Euclid to a Zen Buddhist, whose every utterance is deliberately irrational?'

We are not entitled to reject such assertions just because they are uttered with a hint of cultural superiority. Yet I do not think we need take them seriously into account, for several reasons. As Lévi-Strauss has demonstrated, the supposed credulity and irrationality of primitive peoples is often over-emphasized in anthropological writings. Primitive systems of thought are not illogical. They respect such general principles as 'seeing is believing'; they know the difference between cause and effect; and they make regular categorizations of natural objects. From our point of view, the weakness of primitive thinking is the incorporation of incongruous and (to us) irrelevant elements in the chain of deduction – totem animals, ancestral spirits, ghosts and such like – which can, of course, lead eventually to the most bizarre consequences. But then many Western scientists seem not to be hampered by belief in the Virgin Birth, the infallibility of the Pope, the prophetic foresight of Karl Marx, the unwholesomeness of the flesh of the pig, or the superiority of the American Way of Life. I do not see the harm in a little totemism, white magic, or contemplation of Nirvana – provided it is out of office hours. The initial premises and modes of thought of science are essentially those of matter-of-fact, everyday life, and not those of the sort of metaphysician who feels that he has to prove that the tree is still in the Quad when nobody is about to see it. The primitive savage – *and* the plain man in the streets of London, Paris, Tokyo, or Delhi – has many non-scientific beliefs, which another lifetime of education could scarcely correct; but he is, none the less, a naïve realist at heart. Only a very small proporation of the population, even of the most civilized societies, is susceptible to a genuinely *anti*-scientific, genuinely irrational, metaphysic. It takes a good dose of neurosis, or an altogether too literary education, to arrive at such a decadent intellectual impasse.

We must be careful here not to confuse the conditions required for the growth and spread of science from those required for it to appear spontaneously. However much we may admire the technical achievements of various other civilizations, we must concede that the science about which we are here talking came to birth in Western Europe in the seventeenth century, and has not had the occasion to appear independently at another time and place. One of the grand problems of cultural history is just why it did appear there; what were the origins and causes of this unique event. Of course, the whole thing is very well documented; but it still remains almost as baffling as the corresponding biological problem of the origin of life. This, indeed, is a good analogy, for the conditions envisaged for the original spontaneous generation of the very first living organisms are entirely different from those which could easily be colonized in the later stages of biological evolution. The sophisticated science of today could certainly not now be expected to appear all of a sudden out of the forests of the Amazon or the aboriginal wastes of Australia; but we know that it can be taught quite well, and can thus reproduce itself, in almost any civilized cultural environment.

But although the basic metaphysical foundations of science are relatively naïve, we all know what an immense superstructure of actual knowledge they carry. The unity of science – the fact that it is the product of innumerable hands, cooperating over the centuries – and the unique power of the dialectical processes of imaginative creation and critical reappraisal, make it by far the most elaborate intellectual construction that the world has ever seen. If one is to contribute to this vast system, one must become the master of what is already known, at least in some limited aspect. Occasionally an ignorant scientist may stumble accidentally upon an important new discovery but such serendipity must not become an excuse for lack of education as a policy. Self-taught geniuses, like the incomparable Srinivasa Ramanujan, are entirely exceptional and prove nothing against the necessity for the deliberate teaching of scientific knowledge.

This is of course, fully recognized, and the first condition for the growth of science in a developing country is the establishment of centres of higher education. It is here that the child brought up in an advanced society has his major advantage. Up to the age of ten or twelve he is still no better than a primitive savage, with only a material, instrumental, familiarity with the products of modern technology to offset his natural beliefs in magic and mystery. But the ordinary curriculum of secondary education introduces him rapidly and efficiently to the general principles of scientific thought, and his talents are guided by experienced teachers into the higher realms of knowledge. If he works hard, then he may have learnt enough by his early twenties to begin to do research on his own account.

For a child born into a developing country this process is far more erratic and uncertain. His whole environment may be entirely traditional, materially and intellectually, and it may only be a stroke of fortune that carries him beyond mere literacy. He will be lucky if he encounters good secondary teachers to give him the necessary grounding in scientific subjects, and it is more than likely that most of his university education will be at the feet of scholars who have not, themselves, fully mastered their subjects, and who can convey only the outward inessentials to their pupils. As I have already emphasized, university science teachers who are not in contact with research find great difficulty in keeping their knowledge fresh and up-to-date. The most serious obstacle to the creation of a genuine scientific community in a developing country is often the prior existence of a self-perpetuating academic system of low quality.

This problem is not confined to poor and technologically backward countries where higher education itself is relatively new. In a number of European countries where universities have existed for centuries, a tradition of uncritical book-learning, taught by lazy professors to a mass of students who are only seeking a nominal qualification for a white collar technical job, is the greatest impediment to scientific progress. It is instructive to notice that the best basic research workers in such countries often come through the 'Technical High Schools' or 'Institutes of Technology' which are supposed to be for the training of practical engineers. The reason is that these schools have developed outside of the old medieval university system, and as their qualifications have become more highly regarded they have built up very high standards of competitive entry and rigorous professional training. They make the most of the traditional four to five years' course and give that thorough grounding in the basic sciences which is the hall mark of Continental European scholarship at its best. Whether or not it makes sense to copy such institutions in other countries with different cultural traditions, the value of a few such Centres of Excellence cannot be denied. The recent Indian decision in favour of such a policy can only be welcomed.

But a good basic education in the established principles of mathematics, physics, chemistry or biology is only half the training of a proper scientist. The longer this is drawn out, the more difficult it becomes to learn the other half – how to do research.

The harshness of the psychological transition from being a student to being a researcher is not appreciated in the world at large. Even in scientific circles – especially in the older Oxbridge tradition – it is sometimes assumed that a man who has shown that he has a 'first-class mind' by answering the stereotyped riddles of the Tripos can be put down in a laboratory and safely left to get on with his research on his very own, with no more assistance than is appropriate to a junior colleague. That is what happened in Rutherford's day, so why should we act differently now.

Well, of course, Rutherford was a genius, and had enormous reserves of spiritual energy. The policy of 'throwing 'em into the deep end and larning 'em to swim' is not inappropriate to a strongly independent 'inner-directed' personality. Although I would not go all the way with Feuer in his attempt to show that modern science was founded by 'hedonist libertarians',[11] and I am certainly not convinced that there is a direct intellectual link with the rise of capitalism or of Protestant theology, I am sure that this facet of the scientific attitude comes relatively easily and naturally to persons whose psychology has been moulded by the Puritan culture of northern Europe and the United States. The professional scientist must be able to shut off his ego from the opinions of other people, and 'press on regardless' with his own ideas.[12] In an age when bureaucratic conformism is said to have totally infected the human spirit, he must still assert the stubborn 'It still moves' of Galileo, which echoes Luther's 'I cannot do otherwise'. That same psychological type also arises from another more ancient variety of nonconformism – the ghetto. At its worst, the orthodox Jewish personality is pigheaded, intolerant and argumentative; but these are traits which can become the virtues of strongmindedness, scepticism and intellectuality!

Unfortunately, these qualities are certainly not those that are encouraged by a system of mass education geared to the passing of book-work examinations.[13] Even in the best of circumstances, one may be caught in a psychological trap. To become a scientist one must first master the current consensus, which demands many years of accepting the arguments and opinions of one's teachers without serious question; one must at the same time preserve, or acquire, the self-confidence to reject some of those arguments when, in due course, one uncovers evidence against them. There is no simple way of avoiding this dilemma; we must plunge into the educational system and endeavour to struggle out of it again without too much loss of impetus. In a strange way, some of the most brilliant scientific intellects have escaped the trap by never falling into it in the first place; either arrogantly, or in a dream, they simply opted out of the competition for early academic success, and preserved themselves for their imaginative, critical, creative role as research workers.

Again, we cannot legislate for the Einsteins and Darwins of this world. In the past, when science absorbed an infinitesimal proportion of the manpower and gross national product even of the most advanced countries, one could safely leave training in research to the rough-and-ready competition of 'sink or swim'. Now we need to organize and plan this stage in the creation of efficient scientific cadres with care and forethought.

By convention, this has now become the function of what the Americans have taught us to call a 'graduate school'. Unfortunately the very name – school – tends to put the emphasis on the quality of the advanced courses of instruction offered by the faculty. These are, of course, very important, especially in countries where the ordinary undergraduate curriculum is of a low standard, or too heavily weighted by old-fashioned 'classical' topics. In the ordinary way of things, advanced lectures, both by the resident staff and by visitors from abroad, are the means by which living scientific attitudes and issues are communicated to the students. This is information that cannot be picked up by reading. The textbooks are inevitably a few years out of date, and review articles usually demand too much expertise for the beginner. Although the time of the student should not be overloaded with such courses, in the vain hope of covering all possible aspects of some very large subject, they are absolutely essential, both to those who listen to them and to those who must assemble the material to deliver them.

But a good graduate school is far more important as a training ground in research. This is not just a matter of picking up the practical techniques of the particular science – electronics, glass blowing, computing, plotting data and drawing logical conclusions; much depends upon the quality of the supervisors. For example, one of the attributes of the successful leader of research is his ability to set problems that will yield good doctoral theses for his students and teach them the art of research while making a positive contribution to the advancement of knowledge.[12] It is no good telling a young man to go away and solve the Riddle of the Universe, or find the Philosopher's Stone; unless he is a genius, he will fail, and become hopelessly discouraged; or even go off his head. On the other hand, putting him to work as a mere technical assistant, turning the knobs, reading the meters, and plotting the points in yet another set of measurements with the departmental, white elephant apparatus, will never make him 'self-winding' as a scholar. One of the real disadvantages (besides its enormous cost!) of the very elaborate equipment now deemed essential to research is that a whole team of students and assistants is sometimes needed to mount an experiment. In neutron physics, in high energy physics, in space science and other sophisticated fields, the individual student may get an excellent training in advanced technology, but he is seldom called upon to defeat a real live problem of natural philosophy. Scientists in developing countries may not always be wise to deplore the poverty that cuts them off from these beautiful but extravagant playthings.

Out of this necessity one can scrape a little virtue; the experience of successfully conducting a modest investigation with limited apparatus may well be of greater value to the student than being one of a team of a hundred Ph.D.s manipulating a billion-dollar 'facility' under the distant direction of a powerful boss. Let me give a simple example. Following the lead of several other British universities, we have been experimenting recently in Bristol with research 'projects', in place of set-piece laboratory work in the final undergraduate year. A little problem is set: could the sulphur dioxide emitted by a factory chimney be detected optically from a distance; design an electronic organ capable of imitating any type of musical instrument; what is the actual mechanism by which a steel wire passes through a block of ice? Under the supervision of a member of the staff, two students work together planning experiments, designing apparatus, and interpreting the results. Very little of this work is at all profound, but it has the very important characteristic that nobody – not even the supervisor – knows the answer in advance. In this respect, therefore, these little projects are true to the spirit of science, and provide a most valuable psychological introduction to professional research. The old tradition of a sharp transition from undergraduate course work to postgraduate research may have been quite wrong. The changeover should perhaps be quite gradual, over a period of years; research projects introduced into the undergraduate curriculum are, so to speak, a counterbalance to the further formal instruction now required at the graduate level. In this way, also, we, and our students, early discover their natural inclination – or disinclination – towards a scientific career, and they can move away into more useful and practical professions without loss of face.

But a graduate school is more than a stockpile of scientific learning, technical advice, solvable problems and material equipment. The psycho-logical transition from student to researcher is not merely a matter of strengthening one's personality and throwing away the crutches of book-ish knowledge. Paradoxically, one must also learn to be 'other directed' in a special way: one must acquire the habits and conventions of a respon-sible member of the scientific community.

For example, the graduate student must learn to write scientific papers in the peculiar impersonal style which is now customary. He must learn to accept criticism without personal offence, and to offer it without ran-cour. He must learn to give due credit to other people for their prior discoveries, and not to claim too much for himself. He must learn to scour the literature for references and to keep abreast of all relevant work in his own subject. Above all, he must somehow acquire high standards of scientific accuracy, honesty, and judgement, so as to distinguish quickly between the true and the false, the meaningful and the trivial.

Now it often seems to be taken for granted that these desirable habits of mind and art will come quite naturally to any sufficiently clever young man who is set down to a good scientific problem in a well-equipped laboratory. Nothing could be further from the truth. They do not come from some inner source of soul, but are acquired by imitation of the cur-rent standards and conventions of the particular institution where he is trained. They cannot be mugged up from a textbook – which, in any case, would be about as useful as an instruction leaflet on skiing or horse-riding. It is rightly assumed that the difficult art of research is learnt by apprenticeship, by direct personal contact with an experienced practising master.

This is of the very greatest consequence for the whole theme of my lecture. I have emphasized the world-wide unity of pure science. That unity is not achieved and maintained by such feeble strands as the publi-cation of learned journals, nor by corporate junketings at conferences and congresses; it exists because modern science stems historically from a sin-gle source in Renaissance Europe, and has spread outwards by the natural process of an apostolic succession – by the fact that almost every practising scientist has at one time or another been the personal pupil of a scholar of a previous generation. In our own day we have institutionalized this tra-dition; the regulations for the Ph.D. now practically ensure that one can-not get registered as a professional research worker unless one has been supervised by someone who already has a Ph.D. As one who is often called upon to act as external examiner for Ph.D.s of other universities, I must admit that this ritual is becoming a little like a secular version of the 'laying on of hands' by which a priest is catechized and ordained by a bishop.

The spread of science throughout the world has not been the haphazard distribution of wind-blown seeds, encapsulated within the dry covers of reprints of scientific papers, but by runners and tendrils reaching out from the original institutions and establishing themselves in new soil. Ideas, standards and traditions travel around inside people, and are only transferred from one to another by prolonged contact.

Of course the major developing nations now have self-sustaining scientific communities where this tradition is preserved and retransmitted. But this is quite a recent development, even in some of the most advanced industrial societies.[10] It is interesting to recall, for example, that the St Petersburg Academy, founded by Peter the Great in 1724, was largely staffed by non-Russian scholars until late in the nineteenth century,[14] while even the United States was dependent upon German graduate schools for the training of research workers in many scientific fields until a similar date. The stage of 'take-off' in pure science has been reached only in the past few years in so rich a country as Australia; while small nations, such as New Zealand, or Finland, still find very great difficulties in becoming self-sufficient in graduate studies over a wide range of the sciences.

It is easy enough, at this stage in the argument, to echo the points expressed very properly and concisely by Moravcsik.[15] Graduate schools, he says, should be established in developing countries because of the cost of sending people overseas, because of shortages of places in Western institutions, because graduate students are valuable technical assistants in research, and because once you have sent a man overseas for training it becomes difficult for him to re-adjust to more primitive conditions on his return. All these are valid reasons, and the various suggestions for greater technical assistance by way of buildings, equipment and visiting experts would fulfil genuine needs, which should be given the highest priority.

It is also generally agreed that one should try to give a future research worker the fullest possible preparation in his own country before sending him to an institution for advanced training overseas.[4,16,17] There are many factors, of cost and morale, in favour of such a policy.

Yet there are real dangers in forcing the premature growth of graduate schools before sufficient human resources are available to man them. Only a policy of rigorous selectivity, by the encouragement of really first-rate scholars, make sense. I know how difficult it is to achieve such a policy, against all the academic political arts of log-rolling, me-tooism, appeals to fair play, etc.; but the consequences of flabbiness are all too sadly evident in all quarters of the globe – the proliferation of third-rate research which is just as expensive of money and materials as the best, but does not really satisfy those who carry it out, and adds nothing at all to the world's stock of useful or useless knowledge. The old adage should be stood on its head. Of applied science, it can reasonably be said 'If a thing is worth doing, it is even worth doing badly'; of pure science I am tempted to say 'If a thing is not worth doing, it is only worth doing well!'

The difficulty is that one cannot create good research groups just by setting up graduate schools; the two types of institution are strongly interacting, and come into existence simultaneously. An indigenous school of research is the *culmination* of the growth of pure science in a particular country, not a mere subsidiary agent.

We have got beyond the historical phase where Henry Cavendish or Michael Faraday could work away in his own little laboratory, make his own apparatus, carry out his own experiments and report his own interpretations and theories in communications to the *Philosophical Transactions of the Royal Society*. We find ourselves now with *groups* of professional scientists, organized in teams, departments, divisions, etc., working collectively. If such a group is to live for more than a few years, it must have a balanced composition, with a range of ages and experience and mechanisms for acquiring new members as the older ones retire or become administrators. These are the makings of a graduate school; the problem of tacking graduate students on at the bottom, and arranging for their supervision and formal instruction, is quite trivial by comparison with the problem of creating such a research group in the first place. All active scientific institutions are schools of apprenticeship for their junior members, whether we call them graduate students, research students, 'postdocs' or assistant scientific officers. The responsibility for producing well-formed research workers out of raw graduates should be the most serious burden on any senior scientist or laboratory head. These are the seed corn of the next crop, and the whole quality and success of the scientific activity of one's country depends on how they are treated in these vital years.

Until that happens, there is no substitute for postgraduate and postdoctoral study abroad by the most promising young scientists. Whatever the disadvantages, costs and difficulties of such a policy, they must be borne if a developing country is ever to acquire a worthwhile scientific establishment.

In any case, the notion that a country can ever become so self-sufficient scientifically that individual research workers need not travel is a gross fallacy, only pardonable in a Chancellor of the Exchequer, a cost accountant or a Commissar of Culture. The whole tradition of science feeds on the sacrifices of pilgrimage and exile. Even in Europe, we are learning that polite letters and occasional conferences do not constitute a network of communications; we must actually go and work for a while with our colleagues across the border.

The reasons for this are more fundamental than that 'travel broadens the mind' or that 'one needs the stimulus of a new environment' or even that one can only do good work under the ideal climatic conditions of California – presumably with a good fat Californian income to keep the wolf from the door. The universality of science is at stake.

Consider, for example, the choice of a research problem, and the strategy and tactics of one's attack upon it. The aim is to produce a piece of original, *publishable* research. What this really means is that it must be the sort of research that is of interest to others; in other words, it must seem to contribute to the great body of 'consensible' knowledge.

There is always a stimulating philosophical debate going on concerning the 'importance' or 'significance' of various branches of science, and of particular discoveries in particular fields. It is doubtful whether such questions of taste can be resolved by argument; but it is certain that the vast majority of scientific workers have only a single simple definition of publishable research: it should look as much like other published research as possible, with a few new features to make it seem adequately original.

This remark is not meant to be altogether sarcastic. The great power and philosophical authority of scientific knowledge stems from its cooperative character; each of us builds on and into the work of his predecessors and contemporaries. The rare geniuses are master masons, projecting tremendous extensions of the fabric, but the stones are put in place, one by one, by the numerous journeymen. A piece of research that does, indeed, add just a little to the work of others is more valuable than a disconnected, fruitless speculation, on however grand a theme.

But this means that the topics and techniques of current research are determined by the actions and opinions of the whole scientific community, throughout the world. By its very nature, pure science is completely supernational, for it is essential that every new discovery should be communicated to every scientist whom it might interest, and that it be subject to the critical eye of every competent authority, whether he lives in Tobolsk, Medicine Hat, or the legendary hamlet of Waikikamukau. Everyone knows that science, like love, knows no frontiers; but I do not think it is adequately realized that this is of its very essence. To cut a scholar off from his potential audience (however critical) by administrative, financial, and political curtains and walls is to imprison his mind.

International art fashions or international political movements are signs either of impoverishment of imagination or crude powerseeking. It is sad and unnecessary, for example, that Japanese motor cars should look just like Italian ones, and that Mexican communists should use the same debased Leninist terminology as their Malayan comrades. But there is nothing out of the ordinary in my personal observation that the Mössbauer effect is being studied in Helsinki, and in Western Australia, as well as in its native Germany. Even in China, whose scholars are so unhealthily cut off from direct personal contact with foreign intellectuals, current scientific work is still recognizably international in its appeal.[18] While there may be charming variations of *style* in research from country to country, the idea of a distinctive American, Russian, or Brazilian science, with a different *intellectual* content, obtaining different results, is a contradiction in terms. When, as sometimes happens in wartime, the usual channels of scientific communication and criticism between two countries are blocked, and divergences of opinion develop on purely scientific questions, the very first task of the scholars in both countries is to organize meetings at which the breach can be mended – not by force but by the presentation of the most convincing arguments and evidence until a minimum consensus is re-established. It is the analogy of such meetings that lies behind the Pugwash movement, which has tried to extend the range of consensible topics to include more serious questions of peace, war and politics.

Science in a developing country cannot, therefore, cut itself off from competition, and protect itself by tariffs like an infant manufacturing industry. If it is to be genuine science at all, it must, from the very beginning, be able to stand on its own feet. The international scientific community is not a very lenient examiner, and does not award even one mark out of ten for just attempting the question. Work that is not up to the standard set on the world market of ideas will simply be ignored; it will either not get published, or will appear in an obscure local journal which is not read outside its own country.

Now there is nothing more disheartening and debilitating to a scholar than to know that his work goes unrecognized and unread. His whole professional activity is directed towards the production of some contribution to knowledge; his labour is utterly wasted if that contribution is eventually judged to be negligible. If pure science is to exist at all, in any country, it must be adequate in quality, by this criterion. But only by keeping in close contact with his foreign contemporaries by word of mouth, hand-waving and sketches on the backs of envelopes or lunch table napkins, can the active research worker maintain his critical standards. He must actually meet the authors of the papers he reads; he must have the opportunity to persuade them, face to face, over periods of weeks and months, that he has something of his own to contribute to the discussion. Without such intercourse, his work is liable to drift into a backwater, where he engages in mock debates with himself, quoting only his own papers and ignoring the progress made by other people. Until the day of the universal, free, intercontinental videophone, there is no alternative to carrying the body around when the mind needs to travel.

Even within the range of publishable science with an international audience there is a great deal of choice. The mistake that is often made by those who are a little out of touch is to follow fashion – not even current fashion but the styles of a few years back. This is natural enough, I suppose. Topics and techniques that are heavily emphasized in the literature have a high visibility, even from a distant continent. Imperfect communications will make them seem the only important topics for research; while the peculiar frenzy that excites some ambitious scientists when they see a problem half-way to solution can seem the collective wisdom of the community at large. The phenomenon of fashion in science is of great interest and significance,[19] but its saddest consequence is the tail of ill-equipped research groups who have not quite succeeded in jumping on that gaudy band-wagon as it flashed by. The fact, of course, is that fashionable subjects are those in which the competition is fiercest, and often where too many good people are chasing too few ideas. It does not seem very wise for a small research group, with limited equipment and many distractions and difficulties, to take on the Bell Telephone Laboratories at their own game!

The optimum strategy for the organization and planning of pure scientific research in a developing country would thus seem to be to concentrate on a few solid scientific problems, not necessarily those that are currently fashionable, and to establish a sound reputation for good if unspectacular work in these particular fields. The free trade in knowledge, which is the key to the so-called scientific method, demands specialization and division of labour. It is trite to remark that no single person can now have the whole of knowledge as his province, that the era of the universal scholar is long past. I would go further, and say that the age of the universal university is over – that even such great institutions as MIT and

Imperial College cannot hope to have experts in every field of science on the faculty.

It is much more profitable to emphasize particular lines of research, allowing each research group to reach the critical size for continued viability. I have thrown a spanner at team research which reduces the graduate student to a mere pair of hands; on the other hand, most modern scientists work best if they have a number of close colleagues and contemporaries with whom to discuss problems and from whom to acquire the unwritten expertise of the subject. One cannot lay down an optimum size for such a group, but my own guess is that one needs three or four permanent, established staff, with perhaps a dozen juniors and graduate students, to generate enough intellectual heat, by their mutual interaction, to keep the pot boiling. There are, of course, many forces leading to the fragmentation of research – and there are some gross evils in the building of academic empires – but this degree of concentration and specialization should be the aim of any deliberate policy for the encouragement of pure science, however 'advanced' the country.

Such a policy, however, demands just as much attention to the machinery of communication and travel as trying to follow fashion. It is necessary, for example, to know, and be known to, the few other research groups in the world who are also interested in the same topics. The formal exchange of reprints and preprints is not enough.[15] To ensure coordination, collaboration and fruitful competition rather than sterile shadow boxing, the leaders need to be on terms of personal friendship, and there must be regular exchanges of persons between the groups. I cannot emphasize too strongly the difference that it makes to one's attitude to a paper when the author is not just an outlandish, unpronounceable name, but that clever young man one met last summer in Uppsala – or, conversely, that long-winded old fogey from Nebraska who was mooning about the Department a couple of years ago. Perhaps such emotional prejudices ought not to be, but they are to be reckoned with in the real world.

I am fully aware that this optimum strategy is a counsel of perfection which is not at all easily followed. The reality of pure scientific research in many developing countries is often tragically wasteful of training and talent. There is a regular pattern of failure, which continually frustrates the well-meant effort to construct viable scientific institutions. Let me draw attention to one typical phenomenon.

Let us assume that a young and able student has taken himself off to a good graduate school, in his native country or abroad, and has got his

Ph.D. He has the makings of a competent scientist, and given a decade of experience in active research he might well have become the powerful nucleus of a good new research group. But for perfectly proper reasons – patriotism, availability of jobs, family and cultural ties, etc. – he takes a relatively junior position in a small university away from any major scientific centre. What happens to him?

We do not know in detail, because no sociologist has yet had the inspiration to go and find out. But it does not take all the professional technique of sociology to ask a few questions, and to make a few modest inquiries. On this topic I can strongly recommend an excellent article by Amar Kumar Singh,[20] who discusses very perceptively the impact of foreign study on Indian students. He expresses very forcibly the dissatisfaction and disillusionment with such things as (I quote) 'nepotism and corruption in public organizations and government; poverty and low standards of living; waste and dishonesty; low morality in commerce; red-tape and bureaucratic delay; the discouragement and obstructive attitudes of senior persons in positions of authority; the general inefficiency, lethargy and disorganization permeating all spheres of social and political life, and the absence of social justice and individual dignity'. Stevan Dedijer, in an equally forthright article adds, for good measure, the lack of any general cultural sympathy for the scientific point of view, and remarks on certain pathological forms of careerism that flourish in this unhealthy soil.[21] I have quoted evidence from India, but much the same would be said about many developing countries by those who know them best. One could even apply some of these remarks to a region such as Sicily, where the dishonesty and moral corruption of the Mafia is a formidable obstacle to the scientific spirit.

It is not surprising, in these circumstances, that many a promising young scholar abandons serious science at this stage. He loses enthusiasm for research, and becomes eventually one of those lazy, do-nothing politicking professors at whom so many brickbats are thrown – or else he resigns his post, and flees to a bigger centre at home or abroad, where he can continue to be a scientist. At the moment when he is trying at last to stand on his own feet, he is bowled over by the irresistible tides of an inimical cultural environment. It is asking too much to expect superhuman fortitude in the face of such circumstances.

I see little point in belabouring such a familiar topic and sermonizing on the ills of other people's cultures. One might as well address stern admonitions to the weather, and rage, like King Lear, at the elements! But I do believe that there is a very special factor which particularly affects the professional scientist, and about which something can be done. He becomes intellectually isolated.

As I have emphasized throughout this lecture, the romantic picture of the scientist as a lonely hero, on a solitary expedition through a sort of starlit outer space of the mind, is applicable only to a few extraordinary geniuses; whatever *we* do, *they* look after themselves. The modern research worker is a highly trained professional, who has been taught to cooperate – and compete – with his contemporaries. The better his training, the more he will have to come to depend upon the face-to-face contact with his peers, for discussion, stimulation, criticism and technical information. He is no better prepared for doing research on his own, with no-one near to whom he can speak about his work, than the inhabitant of a city is equipped for life in the middle of the Sahara. If he is to pass safely between the Scylla of scholarly decay and the Charybdis of the brain drain, then we – the international scientific community – must try to help him. We owe this not merely to our own charitable instincts but to our allegiance to the pursuit of learning, for he is not merely a fellow man but a scientific colleague. What can we do to keep his mind alive?

Here again, the only cure is to provide some means of personal contact with other scientists in the same field of study. The admirable practice of sending experts out on lecture tours from the major centres to the provinces, or from advanced countries to less developed regions, helps a little; but it has necessarily only a transitory effect that cannot be repeated often enough, or for long enough on each occasion. It is important to remember that we are not so much concerned with particular pieces of knowledge – the latest formula for the elixir of life, or where to look for quarks, or what would have been found at the bottom of a Mohole – but with recreating an intellectual environment. The visiting expert may convey a great deal of technical information in a brief visit and by criticism or encouragement may decisively alter the direction of research in particular cases. But he cannot transport the atmosphere of a whole institution.

In one of his sympathetic and realistic articles on this whole subject,[5] Moravcsik has made much of what I should describe as the missionary approach to the problem – the sending out of young but competent scientists from the advanced countries to spend several years or more in the newer institutions to help them set high standards of teaching and research. This too, is an admirable activity, but it is not really cheap in money and manpower. It demands, indeed, a missionary spirit of charity and humility which is not always to be found in psychological combination with scientific ability, and it carries with it just a hint of the anxious condescension that is so damaging to the whole missionary enterprise. As I have already remarked, the exchange of scholars for substantial periods is far and away the best means of linking the institutions for which they work, but it needs to be justified on its own merits, and as between near equals.

Another valuable development is the organization of local, national and regional conferences, summer schools, and other meetings. One of the lamentable consequences of intellectual isolation is *isolationism,* the unconscious fear of exposing the inadequacy of one's achievement to one's fellow scientists. In the struggle to get any work done at all, the effort to create time, funds and organization for such meetings begins to seem not worth the possible benefits.

This is a mistake that we have long been making even in Britain, and in Europe generally. The importance of a scientific meeting does not lie solely in the formal communication of the results of research; it is an opportunity for a scientific community to become conscious of itself, and for individual scientists to become aware of one another's difficulties and common needs. Meetings cost money, and they demand public-spirited initiative from those who arrange them, but they could be the saving of many sound scholars.

The above remedies for the chronic malady of intellectual isolation attack the cause, but the disease is so deep-rooted that more drastic cures are needed. A day's visit from a travelling expert, a week at a conference, a fortnight at a summer school – these are only occasional episodes that do not really change the pattern of one's life. Somehow we must make it possible for young scientists in developing countries to maintain prolonged contact with their fellows, over periods of months and years.

The most ambitious project with this goal is Abdus Salam's International Centre for Theoretical Physics, at Trieste. This was created in 1962 as a place to which active scientists from developing countries might go from time to time, for periods ranging up to a year, to bring themselves into contact with current research and with their contemporaries from both developing and advanced nations. One might say that it was conceived as a permanent meeting place for the members of the Invisible College of this particular scientific subject, where they might return again and again to recharge their intellectual batteries. By arranging lengthy advanced courses of study, and a programme of seminars by distinguished visiting scientists, Salam has made it an institution of the very highest standing, permeated with and radiating the best scientific traditions.

The work of the Trieste Centre is so very relevant to the whole theme of this lecture that I wish I had time to say more about it. I should have liked, for example, to speak of the experience of directing one of these advanced courses, which was attended by young research workers from about the most diversified list of countries that you could imagine, and which nevertheless demonstrated the complete universality of the scientific attitude in this particular field. This experience was, indeed, so moving and revealing that it is the real basis for my temerity in giving this lecture; and much of what I am saying here is merely a reflexion of their own feelings about their situation.

The International Centre for Theoretical Physics can deal only with a part of the problems on a very small sector of the whole front line of modern science. I am by no means sure that it is in all respects an ideal solution, nor that the same thing could be done in all other fields. But it has produced one entirely new social device – the scheme of *Associateships*.[22,23]

Briefly, this is a scheme by which an individual scientist acquires the means to travel to Trieste and to spend up to three months there, once a year, over a period of several years. It is like a travelling scholarship, or post-doctoral fellowship, but broken up into small pieces, and spread over a much longer time. Travel and subsistence costs for each visit are paid for, but the associate must, for his part, continue his ordinary duties as a lecturer, professor, or scientific officer in his home country in the intervening periods.

The effectiveness of such an arrangement rests upon its being so well geared to the true needs and time-scale of the isolated scientist. A few months at a time is as much as he can be spared from his official and domestic responsibilities, and yet is long enough to make significant progress on a new scientific problem. The annual repetition of this privilege over, say, five years, allows his work to continue and grow, with opportunities for critical evaluation each year. It stabilizes him in his native environment and yet gives him a seat at the international 'high table'. By assigning an appropriate interval of time to each of those conflicting responsibilities that tend to pull him apart, he can largely reconcile them, and serve each with honour.

But there is no reason at all why such a scheme should be arranged solely in connexion with an international centre of the Trieste type. Every good university is an international centre of learning. The cure for isolation is to come back, from time to time, to any active research centre in one's particular field. The very strength of the scientific community is that it is cross-linked in all directions, and does not depend upon just a few key 'centres'. It is still true, however, that ideas tend to flow outwards, from the most successful and active institutions (not always in advanced countries) to the isolated workers on the periphery. Let the means be found by which the latter can travel regularly to the places whence their ideas come, so that they may share a little in the making of them, and carry them home inside their own heads. The associate thus becomes, so to speak, a foreign member of an active research group, familiar as a person through his regular visits, and a party to its collective wisdom. On the scale on which money is lavished on the support of science, the cost of such a scheme would be negligible by comparison with its benefits. Here is the most practical means by which the scientists of advanced countries can give aid to their colleagues in developing countries – aid that is more precious than shiploads of books and apparatus.

As I come to the end of this lecture, I realize how much more there is to say. I am aware, too, that I have made an elementary error of scientific strategy; I have tackled an insoluble problem, and merely recapitulated all the familiar, trite non-answers to it. But the problems of the growth of pure science in developing countries are not scientific problems, which by definition always have 'answers'. They are by no means always technological or economic problems – although these are large factors in the equations. They are historical, political, cultural and psychological problems – that is, they belong in the realms where there are never 'solutions' but only influences. They demand more than knowledge; they demand wisdom, charity and heroic strength. I bow to those men and women, in all quarters of the globe, who have shown by their scientific achievements under the most adverse circumstances, that these virtues are still attainable.

References

1. Blackett, P. M. S. (1967) *Science* 155, 959–64.
2. Johnson, H. G. (1965) *Minerva* 3, 299–311; (1966) *Minerva* 4, 273–4; (1967) *Minerva* 6, 105–12.
3. Toulmin, S. (1966) *Minerva* 4, 155–69.
4. Kidd, C. V. (1968) in *The Politics of Science* (edited by W. R. Nelson). New York: Oxford University Press.
5. Moravcsik, M. J. (1964) *Minerva* 2, 196–209.
6. Crawford, M. (1966) *Minerva* 4, 170–85.
7. Reagan, M. D. (1967) *Science* 155, 1383–6.
8. Ziman, J. M. (1968) *Public Knowledge*. Cambridge University Press.
9. Odhiambo, T. R. (1967) *Science* 158, 876–81.
10. Basalla, G. (1967) *Science* 156, 611–22.
11. Feuer, L. S. (1963) *The Scientific Intellectual*. New York: Basic Books.

12. Merton, R. K. (1968) *Science* 159, 56–63.
13. Karve, D. D. (1963) *Minerva* 1, 263–84.
14. Vucinich, A. (1963) *Science in Russian Culture: A History to 1860*. Palo Alto: Stanford University Press.
15. Moravcsik, M. J. (1966) *Minerva* 4, 381–90.
16. Maheshwari, P. (1964) *Minerva* 3, 99–113.
17. Blackett, P. M. S. (1964) *British Universities and the Developing Countries*. Convocation address to University of Leeds, 4 July 1964.
18. Orleans, L. A. (1967) *Science* 157, 393–400.
19. Hagstrom, W. O. (1965) *The Scientific Community*, pp. 177–84. New York: Basic Books.
20. Singh, A. K. (1962) *Minerva* 1, 43–53.
21. Dedijer, S. (1963) *Minerva* 2, 61–81.
22. Salam, A. (1966), *Minerva* 4, 461–5.
23. Editorial (1968) Foreign associates. *Science* 161, 421.

29

Three Patterns of Research in Developing Countries*

The creation of a scientific community in a developing country requires shrewd planning over several decades.

The difficulties to be overcome in the creation of self-sustaining indigenous scientific communities in developing countries are legion; material, psychological, political, administrative problems have to be solved simultaneously by a sustained effort over long periods. In the present paper, my purpose is to consider 'research' as a general activity, and to ask why it should be done at all. This will help us to establish principles and criteria for the choice of appropriate fields and topics for research; the establishment of these principles and criteria is perhaps the most critical step in the whole process. It is easier to see what are the material and other practical difficulties and what needs to be done to overcome them; the problem of choice requires a most careful analysis of motives, evaluations and effects which are all too easy to ignore or to take for granted. Failure to think this problem through, and to make a consistent long-term plan of action as a basis for decisions as they arise, may well be one of the gravest weaknesses in the science policy of developing countries.

The question is why do research at all? When we understand the answer to this question, then we understand what sort of research to do. There are three general answers, which correspond to three different time-scales of planning and development. Each of these time-scales is associated with a particular type of research. I shall deal with each time-scale in turn.

The short-term justification: solving practical problems

On a time-scale of a few months to, let us say, five years, research on particular practical problems is of immediate benefit to society. Such topics for *applied research* can arise in any branch of science, although in a

* Paper presented at a meeting of Nigerian physicists, at the University of Ibadan, June 1970; published in *Minerva* 9, 32–7 (1971).

developing country they are not likely to involve the more refined, pure and abstract disciplines, such as elementary particle physics or molecular biology.

This 'Baconian' or 'utilitarian' reason for doing research is powerful and respectable. It draws immediate support from the surrounding society, enhances the prestige of the university in the eyes of practical men, and helps directly in the great labour of social development, not only through the problems which it solves but also by the creation of a corps of practical experts who can bring scientific ideas to bear on industrial or social activities.

In fact, some commentators on science in developing countries have insisted that *only* directly applicable research should be undertaken, if any at all. They make the criterion of immediate social relevance paramount. But there are, as we shall see, other reasons, derived from longer term needs, for active research in fields where these immediate benefits cannot be guaranteed, so that concentration on technical problems which must be and can be solved in the short term would be short-sighted.

What is not, perhaps, fully realized by the 'technophiles' is that it takes an act of imagination to identify practical problems which could be solved scientifically by the use of a few 'academic' scientists. The scientists themselves need to get out of the university laboratory and actively seek such problems, in industries, transport, communications, agriculture, mining, medicine, etc. It is no use sitting around waiting for the practical people in the practical world to come knocking at one's door. The best one can do, in many instances, is to choose a topic which is *potentially* applicable, without knowing precisely what problem requires immediate solution in the hope that, over a longer period, real partnership between the university and industry or government will grow up.

Why do university scientists often spurn applied research? They fear that they will become bogged down in a special local problem which lacks the universality of international science, and hence will lose contact with the world community. The mystique of pure science is to emphasize general problems which are not of immediate practical importance, in order to delve deeper under the surface of things. Precisely this detachment is considered to be a major virtue. But for the scientist in a developing country this aspiration may be aiming too high as an initial goal. If the facilities are poor and if staff are not well-trained, there may be a much better chance of succeeding in relatively modest applied research, before more difficult problems are attacked. When one actually gets into the heart of an investigation, its 'importance' on the international scale becomes irrelevant: the problem itself becomes the immediate interest, and success breeds the self-confidence to tackle a more difficult task the next time. It must also be remembered that the scientific questions that have to be asked and answered in almost any piece of applied research generate further studies of a more general kind. It usually turns out that 'pure' scientific knowledge does not contain the information we need to solve our 'practical' problem, so that we are forced to become more 'basic' as we go more deeply into the 'practical' problem. In other words, an apparently practical and immediate problem, chosen for its utility and as an exercise of technique, may become the starting point of an original investigation of some grand mystery of Nature.

Justification on the 5–20 year time-scale: technical education

The science departments of the universities are major disseminators of advanced techniques in a developing country. They train students in their own specific disciplines, and also provide service courses for students of technological subjects such as engineering and clinical medicine. It is essential that this teaching should be lively, efficient and up to date. Experience has shown that these virtues can be preserved only if the university staff are themselves actively engaged in research, so that they are continually kept aware of the most modern developments in method and principle.

What sort of research is best to fulfil this need? It is obvious that it should involve general techniques of practical utility – for example, electronics, instrument design, computing, chemical and biochemical assaying, etc. An active research school should not only produce finished Ph.D.s, ready to teach in universities; it should also produce the M.Sc. student who moves into industry carrying with him many practical skills which will be needed as development proceeds. Conversely, research that requires outlandish techniques, such as very low temperatures or very high magnetic fields, or the focusing of beams of elementary particles, makes no contribution on this time-scale. Such techniques are not only far too expensive: they are of such limited applicability that it is a gross extravagance to undertake research for which they are necessary tools. In the modern world it is not reasonable to insist that one should do research with sealing wax and string apparatus unless one has the good fortune to discover a new field where important results can be won very cheaply. Nevertheless, there is a very wide range of costs for different types of research, as well as in the diversity of uses to which a particular instrument may be put; universities in developing countries should not enter upon the most expensive types of research.

Another important criterion in the choice of a research topic is the possibility of exploiting special local circumstances such as climate, geography, industry, agriculture, etc. This might take us beyond mere 'applied' problems. Many quite general and pure scientific disciplines demand the detailed study of the local environment, or deal with unusual phenomena that are only to be observed in particular parts of the world – for example, volcanoes, earthquakes and geothermal phenomena in New Zealand, or glaciers in Alaska. It is shrewd to focus research in local universities on the specially significant features of the local environment, thus gaining an advantage over less fortunately located institutions. It is also an obligation on the scientists in that region to devote themselves to the study of such phenomena, as a contribution to the world's scientific knowledge. The fact that such investigations may begin as a quite pure search for knowledge does not stand in the way of their potential applicability as understanding increases, and applicability of direct benefit to the country in which the research is done.

Nevertheless, within the constraints of technical relevance, cost and local circumstances, university research programmes on this time-scale may with legitimacy be on almost any topic which promises to yield results of general scientific interest. The work should, of course, be of publishable international standard – but otherwise there seem to be no precise criteria for choice within the whole range of physics, chemistry, biology or whatever it is. None the less, research within this time-scale can easily go astray.

What happens, of course, is that students go abroad to study for advanced degrees, and bring back with them whatever speciality they happen to have acquired. This is most likely to be a highly 'academic' subject – that is, something directly related to the central themes of the discipline. In physics, for example, our proper emphasis on the fundamental nature of atomic and nuclear physics in undergraduate courses fires the imagination of the students from developing countries who, in their immaturity, have no grasp of the enormous range of special fields where these basic principles are in fact applied. Their knowledge of the science as a whole is so distorted and romantic that they are quite incompetent to make a proper choice at this stage. The power and wisdom to influence such decisions should be exercised quite fully by the scientific departments at which the students take their first degrees, and a lot of discussion of this problem is also needed in the large graduate schools in advanced countries where they go for postgraduate work.

Another big danger to science in developing countries lies in following scientific *fashion*. At any given moment, one finds in the literature a large number of papers on some particular topic – not necessarily of supreme importance in itself, but promising a good return in esteem for the group which manages to solve the current problem which now looks just within reach. This can become a subtle game, in which a number of the cleverest people in the subject challenge one another to a competition, the prize being professional advancement or modest fame. This, of course, is the very worst topic on which to begin research in a developing country; almost certainly, a scientist in such a country will be starting too late to achieve any results before the clever young men in the advanced countries have already solved the problem for themselves, so that his papers fall into that boring and useless category of papers which are referred to in review articles in phrases such as 'Similar results have also been reported by . . . '. If a scientist in a developing country is wise, and fortunate, he will take up a problem of local relevance which is scientifically interesting but not particularly fashionable or highly competitive, and then conduct his research so well that he will build a group which will be recognized as one of the best of its kind in the world. There is no short cut to that goal, but it can be achieved if one does not try to enter into a competitive race for which one does not have the facilities, manpower or atmosphere and in which one will be beaten from the start.

To achieve such excellence it is essential to specialize, and to concentrate intellectual and material resources on one or two research topics in each department. This might seem to contradict the principle that university research should enliven the department's scientific teaching, which has to be spread over many different academic topics within a single discipline. With a staff of a dozen, how can one provide skilled specialists teaching on all these different themes, without reducing each research group to no more than one or two members?

The answer is to insist that one can teach a subject well at undergraduate level without necessarily doing research in that particular subject. The 'academic' themes at the heart of an intellectual discipline are no longer necessarily the most fruitful fields of research, whereas many problems of applied science and technology are interdisciplinary or multidisciplinary, so that an active research team may in fact include expertise over a wide range of techniques and special knowledge cutting across the boundaries of departments and disciplines. For example, a complete physics department could devote itself entirely to glaciology, and yet include experts in nuclear physics (radioactive tracers), electromagnetism (radar sounding), quantum theory (the chemical physics of ice), statistical thermodynamics, solid state physics, classical mechanics, etc. The problem here is to persuade and retrain staff members so that they will collaborate, rather than devising suitable applications of their basic skills. In any case, the challenge of preparing a course of lectures outside the immediate range of his expert knowledge is very good for a young university lecturer, giving him new confidence in his power and a broader grasp of his subject.

The long-term aim: a scientific community

In the long run, university research in a developing country should lead to the establishment of a self-sustaining scientific community, which trains its own members and stands on an equal footing with other such communities in other countries. In the initial stages of economic and social development, scientific research in universities and other institutions may be considered as merely a *product* of social and industrial change, brought into the country from abroad along with higher education and advanced technology. In the end, however, the scientific and technological community becomes a major *source* of economic and cultural development, transforming society with expert knowledge.

The growth of a healthy scientific community in a new environment is not an automatic consequence of spending money on universities. It is essential that this community should have the very highest standards of scientific judgement, and that its members should receive full international recognition. This means that the ablest young men should be encouraged and fostered, and that they should have the experience of active research in a good environment over a long period. In other words, in terms of 20 years or more, research is justified as a means of educating an intellectual élite in the 'scientific attitude'.

This is not an inborn trait, nor is it the product simply of formal training in various scientific techniques or disciplines. It grows by contagion from a few leading research workers who imbue their students with the necessary mixture of imagination and criticism, boldness and meticulousness. It also involves some complicated psychological and social conditioning; the research worker learns to cooperate with others whilst still retaining an independent mind and acquires a peculiar mixture of pride and humility, confidence and scepticism, energy and contemplativeness.

It is very difficult to lay down criteria for the choice of a research field in which these qualities will be encouraged. The apparent 'importance' of the subject in terms of winning a Nobel Prize is quite irrelevant; only the intuition of genius can tell what lines will be fruitful or significant in 20 years' time. The experience of technically excellent research in a pedestrian subject may yet provide the background for some great advance.

Perhaps the best opinion one can offer is that a very promising young man should be given strong backing, and too much obstruction to his own ideas as to what research he should do must be avoided. This is, of course, a most difficult principle to put into practice; it demands discriminating judgement, and greatness of heart in those, more senior, who must carry it out.

There may be some value in emphasizing *cooperation* in research in order to improve the communal spirit amongst scientists. A collection of jealous prima donnas does not provide the right environment for the scientific attitude. It is precisely by exercising his talents within a larger group that the able man develops them fully and makes the most of them, both for himself and for others. A developing country in which small groups of scientists, scattered about in separate universities, have built up a tradition of voluntary cooperation in teaching and research would go a long way forward along this path.

30

Internationalism is not Enough*

A great deal of the responsibility for establishing healthy science in developing countries rests with scientific leaders and institutions in the advanced countries.

The World Plan of Action for the Application of Science and Technology for Developing Countries summarized recently in *Europhysics News* is admirable in sentiment. But, like many such documents prepared in the political sphere, it is all vague aspirations and grand designs, with very little attention to certain apparently trivial realities. Take the following sentence: 'Universities and institutes in developed countries should increase the facilities for promising young scientists from developing countries to spend frequent short periods at such centres while retaining their appointments in their own countries.'

Now how do these 'promising young scientists' arise? They struggle through secondary schools where science is taught as a sort of mysterious doctrine, with little contact with experimental hard fact. They work their way through a university curriculum in physics or chemistry, dutifully learning to describe a Fabry–Perot étalon or to identify the terms in the empirical formula for nuclear binding energy. They are told nothing about why it rains, how a motor car works, nor why glass is brittle; but being very diligent and very intelligent they pass their examinations with honours and win a scholarship to study abroad.

Our contact with such a student begins with a letter from the British Council requesting a place for a graduate student from Paradisia to take a Ph.D. in ebullitiogenesis. Graciously, he is welcomed to the team, and made pretty much at home. He is given a research problem and after three or four years he returns home as a fully qualified ebullitionist. Or, when he arrives, he explains that his real ambition is to be a theoretician and, in due course, he graduates in microcosmology with several joint papers with his distinguished supervisor.

Yet many British physicists know quite a lot about Paradisia and its neighbours. By a curious (but immensely valuable) tradition some of us

* Guest leader in *Physics Bulletin* 24, 274 (1973).

go there regularly as external examiners and science advisers. Some of our colleagues have served there for years as research workers, teachers, engineers, doctors or colonial administrators. Our Paradisian research students talk quite frankly about social and political conditions in their home land.

We know perfectly well that experimental facilities in the physics department of the University of Paradisia are quite inadequate for meaningful research in ebullitiogenesis. We know the fate by intellectual starvation of an isolated lecturer in theoretical microcosmology in a country where even *Physical Review Telegrams* arrives six months late. In fact, our main contribution to the laudable aspiration of the World Plan of Action is usually to put our brilliant former student in touch with a colleague who can give him a permanent academic job in Europe or America.

There is so much that *we* could have done, with so little extra thought. For all its grave difficulties in teaching and research, the physics department of the University of Paradisia is the only source of physics, physicists and teachers of physics in that country. Our example, precept and advice as research supervisors, examiners, science advisers and scientific colleagues can be a decisive influence in its pattern of development.

The first point is, simply, to keep in touch with the reality of phenomena. No doubt our own brightest students can mentally translate the ideal top of the mechanics problem into the rotor of an inertial guidance system or the flywheel of a diesel engine, but the analogy is not so obvious if you have not been brought up in an atmosphere of physical technology. A trained physicist has intellectual and practical techniques that he can deploy, not a head stuffed with the proofs of theorems and historical half-truths. Curriculum reform, new examination techniques, new attitudes to practical work and to projects are under discussion in all our universities; we must export the yeasts of this ferment to our overseas colleagues.

But the key problem is the attitude towards research. So long as university teaching in physics concentrates solely on the famous exploits of Faraday, Einstein and Rutherford, Bohr, Heisenberg and Schrödinger, the students will believe that the only proper aim of research in a physics department is to discover a new fundamental law of Nature, rather than to study the weather, the magnetism of the rocks, jet noise, ion implantation of semiconductors and other such messy subjects. Their desire to do research in ebullitiogenesis or microcosmology is perfectly sincere, but it is based upon ignorance of the alternatives rather than a deep sense of mission to solve the riddle of the spheres. If this ignorance has not been dispelled in their home university, then we must take it upon ourselves to instruct them appropriately.

For there really are innumerable topics for worthwhile physics research in Paradisia. Thunderstorms, the ionosphere, geomagnetism, strange minerals, radio propagation, hydraulics, hydrology, biological structures, archaeological objects, craft processes, buildings, medical devices – the list is endless. To encourage the attitude of mind that only the Kondo effect is worth the salt of a theoretical physicist is a frivolous betrayal of natural philosophy. Of course, *we* know better; but how often do we say to our students and colleagues from developing countries that there is as good science to be found in any aspect of the world of Nature or of technology as in the highly contrived and grossly extravagant pursuit of artificial phenomena at high energies or low temperatures?

Developing countries do indeed need motor mechanics and medical technicians in large numbers; but they cannot do without professors of physics, skilled electronic engineers, science teachers and other sophisticated experts. Every step forward in the World Plan of Action depends upon the availability of people trained to understand, adapt, modify and add to advanced scientific and technical knowledge as it is transferred from developed to developing countries. To build up a viable and properly oriented scientific community in Paradisia will take time; it is almost too late, so we had better start now.

31

Research as if Relevance Mattered*

It is people – the scientists – who will have to translate into action the decisions taken by governments about science and economic development.

The agenda of next year's UN Conference on Science and Technology for Development (UNCSTD) is all intellectual abstractions and institutional frameworks. It has no words for the *people* through whom science and technology will flow. The primary resource for development is skilled engineers, doctors, agriculturalists and other technically qualified people to carry out a multitude of constructive tasks. The training and appropriate employment of such people should be a major theme of the conference.

Technical training is education in what is already known; research is the generation of *new* knowledge. The research scientist is an indispensable person in the development process. But what exactly is his role? The introduction of advanced science into the less developed countries (LDCs) has disappointed many high hopes, especially among the scientists themselves. Not only is their research hobbled by poverty of facilities, geographical isolation, social incomprehension and political harassment; they are uneasily aware that they have a negligible influence on the welfare of their fellow countrymen, and live largely by promises of vague benefits that are unlikely to be fulfilled. This is a tragedy, to which UNCSTD should be attending.

It is abundantly clear, from the experience of the past 20 years in many countries, in many fields of science, that the standard research style of an advanced industrial country cannot be made to work in a country which has a very different pattern of life. This is as true of basic science as of its technological applications. Commitment to the search for truth, and a sincere effort to contribute to the work of the international scientific fraternity, have brought little recognition to LDC scientists, at home or abroad. How is it that so much courageous endeavour, by so many talented people, should achieve so little, for themselves or for others?

A stock response is to blame the economic, technical, administrative or

* Article in *New Scientist* 79, 850–1 (1978).

educational circumstances in which they do their research. These are, indeed, often deplorable. But there seems to be a psychological malaise, a misdirection of motivation, which would not be cured by even the most miraculous draughts of funds, instruments, libraries, technicians, office efficiency, or political liberty.

The LDC scientist has come to accept, without question, the specialized social role of the research worker as a person who makes a 'contribution to human knowledge' within a very narrow field. The traditional ideology of science in the advanced countries encourages competition for personal recognition on the basis of original publications, whose merit is assessed by other scientists for originality, technical excellence, etc. This system works well enough for a large, well-connected community of competent scientists with adequate resources, but fails disastrously on the margins. Those who, for one reason or another, are not in a position to publish work approaching the international standard of quality are simply wasting their time.

To put it harshly, research must be for real; it cannot be cheaply imitated. It is precisely the combination of originality with critical technical skill that makes it a valuable human activity. To mimic the form of this activity, without the substance, is to live a lie, and to mock the truth to which one is supposedly dedicated. No wonder so many LDC scientists are dispirited; no wonder their work is ignored or derided in the international scientific world.

The mystique, the paradigms, the fashions, the triumphs, the limelights of world science exert a tremendous fascination on all research workers. But the essence of research is the elucidation of truths or the solution of problems that *have meaning for the research worker himself.* By referring himself to a very distant group with which he has almost no personal contact, the LDC scientist risks losing all confidence in his own powers, and yet isolates himself from his own local community. Both to repossess his own soul and to make a place for himself in society, he needs to be tackling questions that are significant within that context. 'Relevance' in research implies both social efficacy and psychic commitment by the research worker.

Relevance is a term to be interpreted broadly. It is not just the solution of immediate practical problems. It applies, for example, to basic science that is connected with local geographical, biological or social conditions – on earthquakes in the Andes, for example, or elephants in East Africa, or Polynesian navigational techniques in New Zealand. Indeed, it is precisely by devoting himself to an apparently local question that the LDC scientist has the greatest chance of making the sort of distinctive contribution to world science that he so fervently seeks. Instead of being at the margin of a big, fashionable, highly competitive subject, he may soon find himself at the centre of a less visible but no less scientifically excellent field in which he has proximity to his material.

But of course, nowadays, *everybody* is for relevance in research. In particular, LDC scientists are continually being urged to solve the practical, industrial and agricultural problems of their societies, and are furnished with research grants, or collected into research institutes, to pursue these goals. These problems are close enough at hand and serious enough to be tackled with enthusiasm. Why, then, does applied science also have so little real success in the LDCs?

Once again, it is easy to point to material and administrative deficiencies. Or it is argued that for political reasons the research effort is not applied to the real problems of development, but merely imitates weakly the type of sophisticated technological research that is done in advanced countries. But the underlying issue is the same as in basic science – the morale of the scientists themselves, and their own perception of the problems with which they should be concerned. Judged by the standards of an applied scientist or technologist from a developed country, they lack practicality and common sense; instead of getting down to Earth and using whatever means are available to make an investigation, they try, ineffectually, and inappropriately, to operate on the problem with whatever complicated apparatus they can lay their hands on. They too are obsessed by an image of advanced science, where the use of elaborate and expensive equipment is regarded as the prerequisite of successful research. Because the problem does not have sufficient meaning for the research worker, its solution is subordinated to the technique applied in the attempt.

This psychological phenomenon is so striking that many thoughtful people advocate the deliberate creation of an entirely different science, relevant to the real needs of the developing countries. This proposal, of course, contradicts every philosophical principle and sociological norm about the unity and universality of scientific knowledge. The truth about the natural world is not different in Lima, or Accra, from what is well established in Cambridge, England, or Cambridge, Mass. But with reference to the goals of science, to research programmes, to educational curricula, to the training of research workers, and even, to some extent, to instrumental techniques, alternatives to the advanced science system will have to be found. The *appropriate technology* for development needs a great deal of *appropriate science,* to be discovered and applied by *appropriate research.*

Morale and welfare

These are conjectures and speculations, aroused by what I personally have seen and heard and read about. The whole issue needs thorough analysis and public discussion, to arrive at some wisdom out of a diversity of perceptions and opinions. But let us make no mistake about it; it is the central issue for every theme on the UNCSTD agenda. The research scientist is indispensably involved in the transfer of technology, the planning of research, the application of existing knowledge, building up an infrastructure for R & D, the cure of terrible diseases, the cultivation of agriculture, the growth of industry and every other process of economic and social development. His morale and welfare will be even more important if the LDCs are to seek technological autonomy and self-reliance.

Will the statements from Third World governments to UNCSTD face up to this issue? Will they discuss the problems of scientific education and training in research? Will they be frank about the low pay, poor facilities, bureaucratic bungling and lack of autonomy of their scientific workers? Will they show their awareness of the isolation of their universities and research institutions both from international science and from local actualities? Will there be a deep and searching debate concerning the role of the research scientist in a country without the framework of institutions, motivations and cultural conventions of an industrial or post-industrial society?

Will the delegations of the advanced countries express their concern about the inadequacies of research in the LDCs? Will they search their own souls for their failure to transfer the questing, problem-solving, self-confident, reality-oriented ideology of science to those countries? Will they look again at the research experience given to LDC scientists in the extravagantly instrumented, technically superb, but intellectually fragmented laboratories of the advanced countries, and ask whether there should be an export market for that type of industry? Have they considered the influence of their scientific attitudes, expressed by research virtuosos without any sympathy for or experience of science in developing countries, on the scientific leaders of those countries?

To most of these questions, there can be only very dusty answers. That is why they will not be asked. The situation, in many cases, is too pathetic, too absurd, too tragic, to be revealed in a great public conference. It will seem more prudent to talk of research programmes, information systems, and interdisciplinary projects than about the people who will be asked to carry out these programmes, run these systems, or complete these projects. The waste, the follies, the tragedies will no doubt continue.

32

Ideas move around inside People*

Science, learning and culture diffuse from country to country through personal travel, pilgrimage, and exile.

A very great scientist such as J. D. Bernal contributes to knowledge as much by his presence as by his published writings. People come from all over the world to study under him, to do research with him, and to discuss scientific problems with him.[1] This is a familiar feature of academic life.

The importance of person-to-person communication in the scholarly community is widely recognized; there have even been attempts to measure statistically the part that it plays in the information system of science. But the phrase I have chosen as a title for this lecture suggests something more.[2] The transfer of really valuable knowledge from country to country or from institution to institution cannot be easily achieved by the transport of letters, journals and books; it necessitates the physical movement of human beings. This mechanism for the diffusion of ideas is practised on a large scale, yet the psychological and social conditions for its success have not, so far as I know, been discussed.

Let me go back to an early example. In A.D. 629, a Chinese Buddhist monk, Hsüan-tsang, then aged 26, set out for India. As his biographer explained:

> The Master of the Law realized that each of these doctors (that is, the Chinese Buddhist sects) possessed some outstanding merit. But when he wished to verify their doctrines from the sacred books, he recognized the existence in them of serious discrepancies, so that he no longer knew which system to follow. He then made a vow to travel in the countries of the west, in order to question the wise men on the points which were troubling his mind.

On his way he had various exciting adventures, especially in the dangerous crossing of the Gobi Desert. But he got to India eventually, and

* Bernal Memorial Lecture, delivered at Birkbeck College, London: February 12, 1974.

for about ten years he travelled around, visiting the various holy places of Buddhism. For several years he stayed at Nālandā, the main centre of Indian Buddhist philosophy, and earned a great reputation for his own philosophical teaching. In 645, after 16 years, he returned to China, and settled down quietly in a monastery, devoting the remaining 20 years of his life to translating from Sanskrit the 600 manuscripts that he had brought back with him. From his death in 664 he has been revered and remembered as the founder of the school of Buddhist 'idealist philosophy' in China. The mythical version of his journey told in the amusing Chinese novel *Monkey* is full of demigods, monsters, miracles and battles; but he was a perfectly serious and high-minded intellectual, of considerable significance in the history of Chinese culture.

This story is told at length in a fascinating book, *In the Footsteps of Buddha* by René Grousset, which gives a vivid picture of central Asia, India, and China at that epoch.[3] It also perfectly illustrates my theme. Let us follow the travels of Hsüan-tsang in a little more detail.

In the first place, before he set out he was already fully acquainted with Buddhist philosophy as it was then current in China; he knew what he should look for when he reached India. This initial preparation of the human vehicle is essential if the transfer of knowledge is to be successful.

That is why a student coming from a developing country, such as Peru or Ghana, needs careful preparation in his homeland if he is to get the full benefit of a period of research in an advanced country such as Britain. Without this initial training, he will be very ill-equipped to understand what he has come to learn. Experience has shown that the foreign student who comes too young and green does not know what he really wants, nor what he should take back with him. Despite all the difficulties and obvious perils, we must encourage the creation of graduate schools in the universities of developing countries, to give adequate intellectual maturity – not to mention psychological and social maturity – to students who may later need to study abroad.

But Hsüan-tsang did not get a scholarship to do postgraduate research abroad! In fact, he left China against the express command of the new emperor, T'ai-Tsung, the founder of the T'ang dynasty, who was at that time having difficulty in establishing himself on the throne and in extending Chinese authority into Central Asia. The dangers of the Gobi Desert were greatly increased by his having to make his way across the frontier in secret.

In our own time, such political barriers to travel are sickeningly familiar. In addition to normal obstacles, such as bureaucratic delays in handling passports and visas, the scientific traveller may encounter deliberate restrictions 'rationalized' as follows:[4]

> 'Freedom of migration' of scientists imposes not only a serious material loss on countries which lose their specialists (for their training consumes a not-insignificant fraction of the national income); it has also a moral–aesthetic aspect. A scientist in any country grows up in conditions of stable society, of common work, of exchange of opinions with colleagues. The elaboration and realisation of many scientific problems requires the participation of neighbouring institutes and of a series of industrial enterprises. As a consequence, by deserting his scientific collective, a scientist takes away not only his own personal knowledge but also a part of the general knowledge, a significant fraction of the co-operative scientific achievement. The 'migrating' scientist considers only his own personal interests, setting them above social ones; and by this act he betrays the interests of the society which helped him to become a specialist, which trusted him, and which shared with him its scientific thoughts.

The policy resulting from such attitudes has disastrous consequences. For example, the fact that very few Soviet scientists could travel abroad was a major factor in the ascendancy of Lysenkoism. From about 1940 to 1960, Soviet geneticists had no opportunity to measure this system of ideas against scientific knowledge in other countries. Without occasions for detailed discussion and debate with well-informed foreign experts, it was extremely difficult to criticize and correct the gross follies of Lysenko and his followers.[5]

Similar barriers to travel to and from China in the last 30 years may have similar effects. Without the opportunity to take an active part in the techniques of acupuncture, for example, how can any foreign scientist judge the value of this practice or succeed in explaining how it works? Published scientific work is inadequate as a basis for a real grasp of new concepts and new methods. This applies especially to less tangible matters such as new techniques of social organization. Do we know enough about the Chinese 'barefoot doctors' to make the best use of the same method? It is obvious that many African, Indian and Latin American medical and social workers should be getting personal experience of this new type of health service, which could be of great benefit to their own countries.

However, Hsüan-tsang managed to escape from China, and passed through the small Kingdoms of Central Asia – Turfan, Qarashahr, Kuchā, Samarqand, etc. In this epoch – long before the triumph of Islam –

there were substantial Buddhist influences in this region and 'The Master of the Law' was treated with great honour and respect. But although he discussed philosophy with local monks, he found them essentially 'provincial'. Being on the outer margins of the Indian cultural sphere, they were very unsophisticated in their doctrines.

This, too, is quite familiar. A small and distant scientific community – I have in mind Australia or New Zealand before the Second World War – depends completely on personal contacts with the metropolitan areas. Not being self-sufficient in people or ideas, it must send its advanced students abroad, to acquire the latest knowledge. Teachers often have to be appointed from other countries. Ernest Rutherford, for example, was born in New Zealand, studied physics there and began a little research, but went to Cambridge as a young man. At the age of 27 he became Professor of Physics at McGill University in Canada; but he soon returned to England. These were outlying 'colonies' of European science, without self-sustaining institutions. The best students tend to leave such provincial centres and do not return; the professors who remain lose the incentive to active scholarship and soon get out of touch. This is the characteristic cause of the 'Brain Drain'; the emigrants carry away their personal talents, which are not compensated by a counterflow of scholars to renew the sources of intellectual vitality in the provincial region.

What can be done about this? A conventional practice is to arrange short visits to these isolated communities by active senior scientists, who give a few advanced lectures and enjoy the hospitality offered by the local academics. No doubt this acts as a stimulus, but the effects are ephemeral. The period of contact is too brief for a genuine transfer of knowledge. Such visits are mainly valuable for the personal links that are established, paving the way for closer intellectual cooperation and association.

A more effective and imaginative programme is offered by the International Centre for Theoretical Physics, at Trieste, founded and directed by Professor Abdus Salam, who knew for himself the meaning of isolation in a marginal scientific community in Pakistan. Youngish physicists holding university or government posts in developing countries can come to Trieste for a few months at a time, on several occasions over a period of years, to take part in courses of advanced study or in research. Without losing contact with their own countries they thus keep in touch with the latest scientific ideas, which they carry back home and teach to their own students. This group not only picks up new scientific techniques; they also discuss amongst themselves, and with physicists from advanced countries, the problems of organizing and using science in the Third World

– a much more difficult subject to communicate by books and scientific journals than such trivial exercises as the solution of a Schrödinger equation!

Hsüan-tsang continued his journey until he came to Ghandāra in the province of Peshawar in modern Pakistan. Until about a century before, this had been one of the most civilized areas of Central Asia. It was graced by a culture of high order, stemming from and melding Greek and Indian influences. This Graeco-Buddhist style was to become, in fact, the stereotype of all Buddhist images, from Ceylon to Japan. But in 475 the region had been invaded by the Ephthalite Huns, who ravaged and destroyed the monasteries and great works of art; this was the state of ruin that Hsüan-tsang found.

What happened to the monks of Ghandāra? Many of them must have fled to other parts of India, carrying their doctrines, relics and works of art with them. Was this the mechanism by which the Graeco-Buddhist style seeded itself throughout the East?

We are naturally reminded of the fall of Constantinople in 1453, which sent the scholars of Byzantium into exile in Western Europe, carrying their precious classical manuscripts and the skill to read Greek. A colony of these in Florence was one of the major well-springs of the classical revival of the Renaissance – a source that dried up when the last of the exiles died in 1520.

The greatest event of this kind in modern history was the destruction of German science and scholarship by the Nazis in the 1930s. About 1600 Jewish or politically hostile scientists were driven into exile, eventually to Britain or to the United States. With them they carried the high tradition and superb intellectual style of the German universities, which they planted and cultivated in their new homes.

Consider, for example, Herr Professor Dr Franz Simon, Professor of Physical Chemistry at the Technische Hochschule in Breslau, the pupil and scientific heir of the great Walter Nernst and the leading low temperature physicist in Germany. In 1933, at the age of 40, he left Germany as a refugee to take up some sort of temporary post at Oxford.

Twenty years later, we might have found Professor Sir Francis Simon, Professor of Thermodynamics at the University of Oxford, in conference with such distinguished colleagues as Lord Cherwell and Sir John Cockcroft. In the intervening period, he had created one of the world's leading laboratories of low temperature physics, had played a major part in the development of the atomic bomb, and had conveyed his characteristically German style of thorough, scholarly and critical research to a new generation of British and Commonwealth physicists. It is impossible, as yet, to assess the debt that we owe to that generation of refugee scientists and scholars. As Charles Weiner puts it, the Seminar moved to a new site.[6] They brought with them more than technical skill; in the words of Eugene Rabinowitch 'the greatest role of the European-born scientists was to change the American concept of science: it used to mean invention . . .'[7]

Hsüan-tsang was not, himself, an exile; he came into India as a pilgrim, and made a long pious tour of the sacred places of Buddhism – for example, Bodh-Gayā, where stands the Divine Tree under which Gautama received His miraculous Enlightenment.

Many scholars travel abroad in a similar spirit of pious pilgrimage. The sacred places of the scientist are, perhaps, such places as Cambridge, where Newton once lived and worked, and Rutherford made so many great discoveries. If not adequately prepared, the foreign scholar may come thinking too much of the past, and not enough of present-day intellectual themes and contemporary sources of knowledge. This can be very misleading. A characteristic error for many developing countries in the past 25 years has been undue concentration on nuclear physics and the theory of elementary particles. Students come starry-eyed, full of piety, and tend to see and to take away only what they originally knew about and came to find. They return home with an old-fashioned and unrealistic attitude to research; it is as if they believed that this pilgrimage to the holy places had inspired them with the mystical power to repeat the 'miracles' of Rutherford's great discoveries or to invent incantations of the same magical potency as the equations of Dirac!

Or the visit may be too brief, and inadequate in depth to capture the full spirit of the scientific method. A sad example is J. C. Bose, the first outstanding Indian physicist of modern times, who made important developments in radio transmission in 1895. He came to England and studied for a while under J. J. Thomson, but on his return to Calcutta he took up very eccentric research on measurements of the emotions of plants.

This brings me to a very significant point: it takes a long time to absorb fully a new idea or a style of thought. Hsüan-tsang spent ten years on his pilgrimage round India, stopping for months at a time to study manuscripts or to consult with the local learned men. He stayed for several more years at the great 'university' of Nālandā, disputing, debating and discussing the major issues of philosophy with the monks and hermits who lived in that neighbourhood. Indeed, he became one of the leading Buddhist intellectuals in India, receiving such honours as a personal invitation by the King of Assam to go and teach there.

Thus, Hsüan-tsang did not merely 'learn about' Indian Buddhist philosophy: he *was* an Indian Buddhist philosopher. He had completely assimilated its ideas by himself living with and using it for his own purposes. This seems to be essential if a person is to carry an idea from one cultural sphere to another and to transmit it safely to other people – and it takes a long time. Three years of study for a Ph.D. are seldom enough to make a research scientist out of a graduate student. My own observation is that even an experienced post-doctoral visitor needs at least two years working in our research group before he can, so to speak, think 'Bristol fashion' in solid state theory.

From what I have said it is evident that Buddhist philosophy was the unifying theme of an *intellectual community* in India in the sixth and seventh centuries. What does this imply? The characteristic feature is the possibility of personal exchange, of scholarly mobility throughout the region. We observe a tradition in which people moved about, preached and taught, in one place or another, without serious obstacles. Despite many local sects and styles, there is an overall cultural unity, maintained by the criss-crossing paths of wandering scholars.

An obvious example of such an intellectual community was to be found in Medieval and Renaissance Europe, before the Reformation. Scholars would move from one to another of the great universities, teaching in the common language – Latin – on common subjects with a common background of thought.[8] Thomas Aquinas studied or taught in Naples, Cologne, and Paris. Roger Bacon studied in Oxford and Paris. Duns Scotus taught in Oxford, Paris and Cologne; William of Occam at Oxford, Paris, Avignon and Munich. Remember the greatest of wanderers, Erasmus of Rotterdam; as van Loon puts it:[9]

> He studied in Paris, where, as a poor scholar, he almost died of hunger and cold. He taught in Cambridge. He printed books in Basel. He tried (quite in vain) to carry a spark of enlightenment into that stronghold of orthodox bigotry, the far-famed University of Louvain. He spent much of his time in London, and took the degree of Doctor of Divinity in the University of Turin. He was familiar with the Grand Canal of Venice, and cursed as familiarly about the terrible roads of Zeeland as those of Lombardy. The sky, the parks, the walls and the libraries of Rome made such a profound impression upon him that even the waters of Lethe could not wash the Holy City out of his memory. He was offered a liberal pension if he would only move to Venice, and whenever a new university was opened he was sure to be honoured with a call

to whatever chair he wanted to take, or to no chair at all, provided he would grace the Campus with his occasional presence.

Or think of the German universities in the nineteenth century. Students and professors would move from centre to centre, seeking special knowledge, or would study together in some famous institution, such as Liebig's chemical laboratory at Giessen. Many of Liebig's students were foreign: between 1830 and 1850 about 60 British and 20 French chemists were trained in research under his supervision. The academic community of Germany fed ideas and techniques, through such students, to other regions of the western cultural area.

Nowadays, the largest intellectual community in the world is that of the United States. The 'Academic Market Place' stretches from Boston to San Diego, from Florida to Seattle. Scientists move freely from job to job, from university to university, from laboratory to laboratory, over the whole of this vast territory. And of course the affluence and intellectual vitality of American science draws in large numbers of Europeans, Asians, Latin Americans and other foreign nationals.

Within this great community the rapid circulation of people and ideas creates uniformity of cultural style and uniformity of scholarly techniques. The stylistic homogeneity generated by this high rate of transfer and diffusion may, indeed, be a source of weakness. We may ask, for example, whether all science must necessarily be done with very expensive apparatus, in a spirit of competitive urgency. There are dangers in a situation where credible alternative styles are not powerful enough to be taken seriously. The arrogance of cultural imperialism can be intolerable and eventually self-destructive. It is absolutely essential to preserve free speech, scholarly mobility, and open competition for academic jobs over the widest possible area if these dangers are to be avoided.

Is the whole world now a single academic market place? The scientific *literature* does, indeed, diffuse to all corners of the globe (although very slowly to the more distant and poverty-stricken lands), but *job* mobility is quite restricted in practice. Political barriers such as the Iron Curtain close off some regions. The language barriers in Europe seem more restrictive now than they were in the Middle Ages. How few French and Italian scientists hold permanent posts in Britain – and vice versa! A few years ago there was a conspicuously greater movement of European scholars to and from the United States than between neighbouring European countries. Scholarship and exchange schemes have encouraged intra-European mobility, but are inadequate in volume and method. One can see the great importance of CERN, the Laue–Langevin Institute at Grenoble, and other deliberately international scientific institutions. The European scholarly community is being created and integrated by the experience of individual scientists working together for many years, in the same building, in the same teams, studying the same problems, using the same apparatus.

On the world scale – for example, as between Japan and Nigeria, or between Australia and Czechoslovakia – the actual contacts may be very weak. Of course, we have scientific conferences. Those early meetings, such as the Solvay conferences, where a couple of dozen earnest scientists would come together from year to year to discuss their common problems, were the progenitors of a vast and expensive industry. At a modern scientific congress, hundreds, thousands of professional experts assemble in great lecture halls, and throw their specialized knowledge at each other in gristly gobbets. People come all puffed up with their own ideas, expose them briefly to the audience, and take them home again without much change. The time available is too short, the programme is too crowded and disjointed, the contact of mind with mind too chaotic and fleeting, to achieve significant transfer. The main benefits of scientific conferences are in the initiation and maintenance of social contacts with scientific colleagues around the world, leading eventually to longer and more intimate occasions for intellectual association and contagion.

No doubt a conference can be a grand ceremonial activity. In 643, Hsüan-tsang was graciously invited by Harsha, the Emperor of Northern India, to visit his court at Kanauj. A great assembly of princes, monks, sages, hermits, etc. was gathered together. There were lots of elephants, pearls, golden bowls and rich presents. Hsüan-tsang then held forth publicly in favour of his particular doctrine – the Mahayāna or Greater Vehicle of Buddhism. The King proclaimed that 'if anybody found a single erroneous word in it, and could show himself capable of refuting it he would have his own head cut off in token of his gratitude'. Nobody did!

The intellectual fruits of this conference were not very impressive. Hsüan-tsang's adversaries – especially the Hindus and Jains – plotted against him and stirred up trouble. A shrine set up by Harsha was set on fire by a criminal hand, and the King himself almost fell victim to a stabbing attack, apparently instigated by the Brahmins. This conference, although very extravagant and spectacular, did not ward off the coming triumph of Hinduism in India. Harsha himself lost his throne only four years later.

The Indian monks and princes wanted Hsüan-tsang to stay on. They said:

> India has seen the birth of Buddha, and although He has left the earth His sacred traces still remain here. To visit them one after another, to adore them, and to sing His praises is the way to bring happiness to your life. Why come here to leave us all of a sudden? Besides, China is a land of barbarians. They have a scorn for monks and for faith. That is why Buddha did not will to be born there. The inhabitants' views are narrow and their sins go deep. That is why the sages and saints of India have not gone there.

But Hsüan-tsang was not very happy. He replied:

> Buddha set forth His teaching that it might spread everywhere. What kind of a man is he who would slake his own thirst with it and neglect those who had not yet received it?

The truth was, also, that he despised many features of Indian life, such as political anarchy and religious excess. At heart he remained a Chinese 'Confucian':

> In our kingdom the magistrates are serious and the laws are observed with respect . . . humanity and justice are esteemed, and old men and sages are given the first place. That is not all – science holds no mystery for them. Their perspicacity equals that of the Spirits; the heavens are a model for them, and they can calculate the movements of the Five Heavenly Bodies . . .

In other words, he was homesick. This is an inevitable feature of the diffusion of knowledge: people must go away from home, to very alien surroundings, for years on end. One remembers, for example, those British teachers who went to the far corners of the Empire, a century ago, founding colleges and universities. The romance of an exotic culture soon wears thin, and the discomforts and social isolation of the expatriate become a burden. To stay on for 20 years to create a viable institution was self-imposed exile.

The case of Japan in the late nineteenth century is typical. In addition to the Japanese students who studied abroad – for example, at German chemical laboratories – a relatively small number of foreigners were employed as advisers and teachers. Henry Dyer went to Japan from Glasgow, accompanied by eight subordinate teachers and assistants; they created a school of engineering which later became the engineering department of the Imperial University and the foundation of Japan's modern technical development.

Even today, when every city seems the same, the cultural contradictions of a foreign country are not easy to comprehend or accept. Here are the words of a student from Singapore:

> Being a Christian I go to a church nearby. I was quite surprised during the first few weeks in this land to find that contrary to a widely accepted belief, this is not a Christian country. Though its ethical standards and public opinion are based on Christian ideals, most of the people do not seem to go to church or practise their faith! Some of the churches which I have visited are almost empty. However, the one to which I am now attached is a vital and flourishing church. Various aspects of life in this country are quite different from that of mine. Attitude towards sex-freedom is unacceptable by our people. Keeping long hair is considered as a reflection of hippieism in my country while it is so common among the young people here including devout Christians.

Many of the Europeans who went to the USA during the past 40 years have not been entirely happy there. As Theodore Adorno remarked:[10] ' "Adjustment" to exile is not necessarily a sign of "maturity" '; he found American sociology not to his taste, and returned to Europe after the War. One of my best friends, a Spanish physicist who was once my student at Cambridge, spent some years at a distinguished American university, where he could have remained for the rest of his career. But, not wishing to bring up his family in that cultural environment, he returned to Spain, to the many disadvantages of a weak scientific community and a hateful system of government. Like Hsüan-tsang, he felt a moral duty to return to his own country:

> China is separated from India by a vast distance and was very late in learning of the Law of Buddha. Though she may have a superficial knowledge of it she cannot embrace it in its entirety. That is why I am come to be instructed in foreign countries. If I desire today to return there, it is because the sages of my country are sighing for my presence and are summoning me with all their prayers.

The return to China was made in much grander style than the outward journey. Not only did Hsüan-tsang take back with him many Indian manuscripts; he also had many Indian carvings, which were perhaps of considerable influence on the beautiful Chinese sculpture of the T'ang dynasty. The Emperor welcomed him in great state, and he settled down to his translations – not entirely forgetting the excitements of his journey for, as his biographer reports: 'He would also discourse frequently with the monks on the subjects of the sages and saints of India, the systems of the various schools, and the distant voyages of his youth.'

In his translations, Hsüan-tsang performed, as always, his precise historical task. He was very much the faithful commentator on the Indian masters, creating a school of Buddhist 'idealist' philosophy in China, on the model of what he had learnt at Nālandā. This is another fundamental link in the chain: ideas that have been carried half-way round the world must be planted very firmly if they are to survive in a new environment. This was, for example, the role played by J. R. Oppenheimer before he got mixed up with making atomic bombs. As a young man he went to Göttingen, and worked with Max Born; on his return to the United States in the early thirties he set up a graduate school in California and by his brilliant teaching transplanted the new style of theoretical physics from Germany to the USA.

For this reason, the 'reverse brain drain', however meagre in quantity, may be a historical force of great significance. People who come home after five or ten years in a foreign country carry with them, fully assimilated, the ideas and inner meaning of another culture. The Indian physicist, say, who has been on the staff of the Bell Telephone Laboratories, or who has taught for ten years at a British university, can play a very important part in the modernization of science in India, because he knows exactly how the American or British system of education and research really works. He may not find it easy to interpret his own understanding to those whom he must influence in his own country, but he has had time enough to grasp the reality that underlies the superficial experience of the foreign student who comes merely to take a Ph.D. This has been a most important factor in the rapid improvement of Indian science in the past decade, breaking away from the weak 'homespun' science of the older professors who had studied only briefly abroad. The same phenomenon is to be observed all over the world. For example, in Brazil, the University of Campinas, near São Paulo, is staffed by a group of Brazilian scientists who have come back from the USA together, and who are setting up laboratories in the US style, to do the American style of research. Similarly the Korea Institute for Science and Technology in Seoul is very dynamic and well-organized, having brought back more than half its staff from abroad.

This brings us to the end of Hsüan-tsang's journey, and life, which so perfectly accomplished its historic mission. But we must not suppose that the life of the mind is all travel and tourism – whether by camel or supersonic jet. Ideas have their own independent means of transport, and can pass through surprisingly dense obstacles. For example, in the early nineteenth century, the Japanese government allowed a few Western books, especially in Dutch, to be translated; there is a charming account by a

Japanese physician, of how he confirmed the accuracy of a book on electricity that he was translating by repeating Franklin's kite experiment, using a tall pine tree.[11]

And ideas must have their origins before they can travel. Just for having new ideas, perhaps it is best to stay at home and think. Diogenes, they say, lived in a barrel. Immanuel Kant was born in 1724, in Königsberg, where he studied, taught and spent all his life. He lived very simply and never travelled; yet he was one of the greatest and most original of philosophers, who sent hundreds of Englishmen and Americans buzzing around Germany all through the nineteenth century.

Or think of Niels Bohr, the deepest and best-loved theoretical physicist after Einstein. For 77 years his home was Copenhagen. True, he spent a year or two at Cambridge and at Manchester, but after that you would find him at the Institute of Theoretical Physics they built for him in Copenhagen. Others came and discussed with him and with each other; Bohr's Institute was the real centre of quantum physics, in the 1920s and 1930s. Very late in the War, in 1944, he was secretly transported to Britain and the USA to take part in the atomic bomb project; Churchill found him so puzzling that he thought he might be a spy! So, at the end of the War, he went home to his Institute – on his bicycle – and stayed there.

I hesitate to suggest that J. D. Bernal was a stay-at-home. He travelled widely, and was widely admired and honoured. It is significant, nevertheless, that his scholarly career was concentrated in London and in nearby Cambridge. The ideas moved around inside his own head, and came out in those brilliantly original and novel forms that we remember and honour on this occasion.

We may recall, perhaps, those Indian hermits and sages whom Hsüan-tsang visited; theirs, too, was a perfect form of the scholarly life, with its own characteristic destiny.

References

1. Miss Anita Rimel informs me that 34 scientists from 17 foreign countries worked with Bernal at Birkbeck College, from 1945 until his death.
2. It comes, in fact, from my old friend, Professor Jasper Rose, of the University of California at Santa Cruz.
3. Translated from the French by Mariette Lyon (London: Routledge, 1932). I am most grateful to Mr F. H. Morgan for telling me of this book and lending me his copy, from which come all the quotations concerning Hsüan-tsang in this lecture.
4. This comes from a piece entitled 'The Alma Mater of a Scientist' by Academician I. I. Mints of the USSR Academy of Science: *Literaturnaya Gazeta,* November 7,

1973. The translation, by Professor D. B. Spalding, appeared in the News Bulletin of the Scientists' Committee of the Israel Public Council for Soviet Jewry: January 31, 1974.

5. *The Rise and Fall of T. D. Lysenko* (New York: Columbia University Press, 1969) and *The Medvedev Papers* (London: Macmillan, 1971), by Zhores A. Medvedev, provide complete documentation on this case.
6. 'A new site for the Seminar: the refugees and American physics in the Thirties' by Charles Weiner in *The Intellectual Migration*, edited by D. Fleming and B. Bailin (Cambridge, Mass.: Harvard University Press, 1969).
7. In *Illustrious Immigrants* by Laura Fermi (Chicago University Press, 1968), which treats the whole subject very well.
8. This is brought out in a paper by Stevan Dedijer 'Past Brain Gain Policies: A Historical Divertissement' in *Cahiers d'Histoire Mondiale* 10, 635 (1967).
9. H. W. van Loon, *The Liberation of Mankind* (London: Harrap, 1926).
10. In *The Intellectual Migration;* see endnote 6.
11. Quoted by G. Basalla, 'The Spread of Western Science', *Science* 156, 611 (1967).

PART SIX

RELATIONS WITH SOVIET SCIENCE

33

Letter to an Imaginary Soviet Scientist*

Obstacles to communication between Russian and 'Western' scientists are harmful to both parties, and should be discussed frankly.

Dear colleague – The news of your election to the Soviet Academy has just reached us here, and I hasten to add my congratulations to those of your other friends and admirers. It does not seem 15 years since we began to notice your name as the author of a series of brilliant contributions to our subject, and wondered whether you were as clever in conversation as you seemed in print. Nor does it seem as long as seven years since we actually met, and I learned to value your cheerful, charming and provocative companionship. You deserve your honour, my dear friend, not only for your scientific work but for these human qualities which we have all appreciated on those few occasions when we have been able to have you with us in person.

Now that you have risen to this august rank you will, of course, have heavier administrative responsibilities – but you will also be able to play a larger part in the general direction of science in your country. Knowing you as I do, I can only feel that this will be beneficial both to the Soviet Union and to our subject throughout the world. The combination, within your person, of intellectual power and human judgement fits you almost uniquely for such a role.

Perhaps, therefore, you will excuse me, even on this especially happy occasion, if I mention something that has been worrying me from time to time and that has cast some small shadows over our friendship. It is not a matter that should stand between us personally – but I hope that you may exert your new-found authority to do something about it.

I allude to the extraordinary deficiencies, amounting sometimes to downright discourtesy, that hinder written communications and practical arrangements between Soviet scientists and ourselves. You must, I am sure, know the sort of thing I mean.

For example, a mutual friend arranged a conference here last December.

* Published in *Nature* 217, 123–4 (1968).

Knowing the problems of 'getting any Russians', he set to work in good time – or so he thought. In January he wrote to his opposite number in Moscow, an Academician, inviting him to come as a guest speaker. He also asked for the names of other good Russian scientists as possible invited speakers and participants. To this letter he received no answer. At the end of March, having nearly finalized the programme of review lectures, he wrote again to the Academician. Still no answer.

At this stage, he would have been justified in dropping all Russian names from the programme. However, being experienced in such business and genuinely wishing to hear at first hand about the work of the Moscow Institute, he wrote directly to the Presidium of the Academy and to the Ministry of Education, officially inviting six leading Russian scholars by name. This letter, sent in May, was answered in October, as follows: 'Dear Professor X, the Soviet delegation to your conference will consist of Y and Z. Please make arrangements for them to receive visas to enter your country.'

Unfortunately, the two names mentioned in this brief and uncivil communication were not among those asked for. Indeed they were practically unknown to our friend, who could only assume that they were junior workers who had not yet risen to international prominence. He wrote back asking for more information, pointing out that he would not now be able to include these names in the official programme as invited speakers, but would welcome them nevertheless. He also set in motion the Foreign Office machinery for granting visas.

Again nothing seemed to happen until three days before the start of the conference, when a cable (in Russian) demanded accommodation for eight Russian scientists, including now the Academician himself. With tremendous energy, the organizing secretary wheedled six more rooms out of the local hotels, aroused the Foreign Office to action, and rearranged the programme. He even set graduate students awaiting at the airport to guide the party. But nobody came.

Finally, in the last two days of the meeting three Russians appeared, unannounced. One of them did not seem very familiar with the actual subject, but they all insisted on reading out their unscheduled communications. In the end, beer and vodka in the local pub resolved many inhibitions, and we soon became good friends. They were, indeed, full of apologies for any slight difficulty that may have been caused by the inefficiency of their bureaucracy – and once again we forgave them because, poor souls, it was not their fault that we had been put to so much trouble.

What puzzles me about such buffoonery (believe me, I do not exaggerate the sort of thing that often happens) is that it benefits nobody. If attendance at scientific conferences is valuable, then surely the most suitable Russian participants should have been chosen, by experts such as the Academician, and efficient arrangements made for their travel. Much expense and effort seems to have been wasted. And the international standing of Soviet science was not enhanced.

Well, of course, we too have our difficulties sometimes in persuading financial committees to make grants for travel to conferences, and in rearranging our university duties so as to get leave of absence; sometimes even nine months is an insufficient gestation period for such important administrative decisions. But at least the advice of expert scientists is followed. I know that it would not be fair to blame your Academician personally for this particular fiasco; but I cannot think of anyone else with the knowledge needed to take such decisions and to bear the responsibility for their success.

You must yourself have suffered through similar events, and do not need to be told by me that they are harmful to the Soviet Union and its scientific development. I can only say, as an outsider, that they really cause us a great deal of trouble, to the extent that many of us, in organizing international scientific meetings, may begin to feel that it is not worth the effort to try to get Russian participation. There is a limit, in terms of time and worry, to the price that we are prepared to pay for this commodity, however valuable it may be to us. I hope therefore that, in your new rank, you will press, through the Academy and other organs of the Soviet scientific community, for a great improvement in the practical management of all such matters.

Yet there is a more subtle aspect of relations between Russian and 'western' scientists that cannot be improved by mere administrative reorganization. I have mentioned letters that were not answered, nay, not even acknowledged. Friends have told me of the rebuffs they have received in planning a visit to the Soviet Union. For example, polite requests to talk to such and such a scientist have been fobbed off or simply ignored. We all know of the custom of translating scholarly works into Russian, without consulting the author, despite the advantages that may be gained from his advice concerning printing errors. Are these and other incivilities always to be forgiven?

Can it be that you and your colleagues are simply unfamiliar with our ways, just as we must be with yours? Consider, for example, the general tone in which correspondence is conducted between established scientists in different countries. It is hard to characterize this tone specifically, but it is always courteous and friendly, whether formal or informal. From an unknown German or Italian, a letter might be written in the old-fashioned style of conventional respect; from an American friend it might be all on first-name terms, with chat about each other's family. In every case, the letter would be addressed as from person to person, without passing through the bureaux of organizations, academies, universities, or other institutions. The courtesies would be those normal between independent individuals, asking perhaps for advice or assistance, regretting in return the inability to give such help, conveying opinions or information, making practical arrangements, etc.

The point is that, however much we may be in fact employees of universities, government laboratories or industrial corporations, we still like to regard ourselves as persons of independent standing, capable of deciding for ourselves on all matters associated with our professional activities – what subjects to work on, when and what to publish, whom to meet and talk to, where to go for conferences and advanced instruction. I do not say that we are in fact quite as free as we like in all such questions; we must, indeed, perform our university duties and otherwise satisfy our directors of research and governing boards – but we are not mere functionaries. To fail to answer a letter from one such scholar to another is therefore something worse than administrative inefficiency – it is a personal insult. To act as if one scientist is interchangeable with another as the speaker in a conference is an attack on the personal standing of each one of us. To give meaning to our work, we need to preserve the illusion that we are contributing individually to the growth of knowledge and that each one of us deserves to be treated as the independent proprietor and cultivator of his own little cabbage patch in the land of learning.

You yourself, I know, have never offended in this way; all I ask now is that you try to explain to your colleagues the importance of such apparently small issues. Our mutual good relations depend on an implicit respect for each other's position by an explicit regard for his social conventions. An improvement at this level would be of far greater benefit than any number of cultural exchange agreements between our respective governments.

These benefits would not be merely a vague increase in 'international understanding'. It is now clearly established, by those who have studied the sociology of science, that the informal channels of communication between scholars are quite as important as the official organs of scientific publication. Information about new discoveries and new techniques diffuses very rapidly through our scholarly communities by means of letters, visits and personal contacts at conferences. To be cut off from such contacts is one of the main complaints of scientists in small countries, far from Europe and North America; it hampers their research at every level, seeming to make it always a little out of date. If we could unblock these channels between the Soviet scientific community and our own, then we should both benefit greatly; we should not have to learn, by painful experience, some of the things that you already know; your research would always start from the best possible basis of knowledge, including some of those subtle ideas that we have already conceived. I cannot emphasize this point too strongly to you, to your professional Russian colleagues, and to all those who have anything to do with the scientific development of your country. I know very well that there are obstacles to the opening of all doors between your scholarly community and our own; at least a general amelioration of the level of civility in personal communications would unlatch a few windows between us.

You will, I trust, forgive me for having spoken so bluntly. I have not the least complaint against the way that you have treated me over these many years. The role of the 'candid friend' is dangerous and foolish in personal relations. But this is not, in the end, a question of the relationship between private individuals; it concerns whole communities, great nations and our common dedication to the pursuit of knowledge; to fail to speak our minds honestly is to betray our social responsibilities. Perhaps we too are unwittingly offending you by our behaviour; give us the opportunity to show our emotional maturity by accepting your criticism in a humble spirit.

Meanwhile, dear friend, we must make the best of our little lives within our chosen spheres. Once more let me congratulate you on the recognition that has crowned those past years of toil and effort and wish you many years of further success, the delights of yet more discoveries, and the fruits of wisdom and understanding.

Arrivederci: to the time when we meet again.

Yours ever,

John Ziman

34

A Second Letter to an Imaginary Soviet Scientist*

How can we personally show our sympathy for scientific colleagues in other countries who are victims of injustice?

Dear Colleague – Your kind invitation to attend the International Conference in Moscow this summer arrived this morning, and I hasten to reply. The programme is unusually interesting. Yet I am not quite sure whether I shall, in fact, accept the invitation. Something troubles me, and I need your advice as an old friend and colleague.

Let me say, first of all, that I have long wished to visit the Soviet Union, not only for its tourist attractions but in order to make closer contact with Soviet scientists in our field. In an earlier letter I remarked on some of the confusions that arise when arrangements are made for Soviet scientists to attend conferences in other countries. If you cannot easily come to us, then we should make special efforts to attend conferences and visit laboratories in the Soviet Union so as to keep each other well informed on our scientific progress.

In particular, I was looking forward to a long visit to your theoretical group, whose head has done such brilliant and exciting work over the last ten years. Imagine my distress, then, to see his name in the newspaper, and to receive from him, by a roundabout route, a heart-rending plea for help and support in terrible circumstances. You know much more about it than I do, for it must have hit you particularly hard that your young colleague, with whom you have worked so closely over so many years, should be dismissed from his post, and from all scientific work, for daring merely to ask that he might be allowed to emigrate to Israel. You and I, as scientists, can feel the additional cruelty of the order that his books – the fruits of half a lifetime of effort – should be withdrawn from the libraries and that his papers should not be cited in future.

I understand perfectly that there is very little that you and your colleagues can do in the face of such forces. No doubt you are helping our

* Published in *Nature* 243, 489 (1973).

friend in private, with the gifts and sympathy so characteristic of Soviet intellectuals. I suspect that you will also devise ways by which he can continue to do research in private and will be sure to keep his name in the scientific literature; no scientist who is remotely capable of judging the relevance of a citation would be an accomplice to this utter perversion of the system on which we all depend for our scholarly recognition.

If Soviet scientists are to be denied the fruits of their honest labours, then Soviet science will soon be at the mercy of dishonest timeservers. You must be trembling inwardly with the thought that a higher patriotism than complete obedience to these orders would be to preserve for the Soviet Union the sceptical, critical, imaginative scientific way of life.

But my real question to you is – what should we do; what should I do, to help the scientific friend whom I have never met? I do not think I am greatly influenced in this by my own Jewish birth and upbringing, which in a free society I need neither exalt nor repudiate. A man in trouble has called on me by name, invoking the universalism of science, and the fellowship of our Invisible College, which knows no natural frontiers; I must surely heed his call.

One form of help is very easy to get in this case – publicity. All normal procedures of protest are being actively pursued, in the Press, in government circles, through embassies, and so on. It is scarcely to be believed that much more can now be accomplished in this way.

Can anything be done through our scientific organizations and learned societies? It is my belief that this particular case – like a number of other cases in various countries (for example, in recent years, Argentina and Czechoslovakia) – constitutes such an infringement of the norms of the scientific community that it is our duty to take notice of it officially. It seems to me, for example, that the agreements made between the Royal Society and the Soviet Academy of Sciences for scientific exchanges are based upon the assumption that these bodies have a similar respect for scholarly integrity, for scientific authority and for the acknowledgment of scientific originality and priority. It would not, I believe, be a transgression of its principle of avoiding 'politics' if the Royal Society were to question the official attitude of the Soviet Academy to these matters, and to indicate its unwillingness to cooperate with an organization that so openly failed to live up to the high ideals that we all profess to hold.

In the end, however, only a personal gesture is likely to be possible. That is why I am troubled about whether or not to accept your invitation to the Moscow conference. Should I refuse to take part myself in an activity sponsored by a government whose policy I detest, under the auspices of an organization – the Soviet Academy – whose hypocrisy I deplore? I realize that such action is largely symbolic. At some cost to myself (for it really would be pleasant and scientifically profitable to come) I cause a little damage to those whom I wish to influence and fail in my friendly duty to people, like yourself, whom I like and trust. Yet I think that the gesture is not quite useless, and that if it were copied by the many hundreds of British, European and American scientists who have been invited to scientific conferences in the Soviet Union this summer it might bring out very clearly the extent to which, as a community, we are united in detestation of these present actions by your government.

Tell me, then, colleague and friend, what I should do. I have never agreed with those who demand a boycott on visits to this or that country – Spain, Greece, South Africa, Czechoslovakia – on general political grounds. A scientific visit may look like support for the policy of the government in power; but in reality, it may be valuable moral backing for those men of integrity and good will who are struggling to establish scientific standards under very adverse circumstances. I prefer to put my trust in such men, and to accept their personal invitations in the name of scientific fellowship.

Is this what you had in mind when you sent me this invitation? Would I come as an official guest of your government, expected to speak in admiration of your scientific achievements and never to express, even in private, my inner feelings on this tragedy? Or do you ask me as a personal scientific friend, not only to discuss our common interest in certain technical questions of scientific research, but also to share our common concern for what is being done to another friend by this cruel world in which we all have to live? I am not speaking, of course, of reckless public acts of defiance, or any other such extravagant, useless gestures: merely of the spirit with which you would yourself meet me on this occasion. Until I know that, allow me to reserve my reply. I hope that you will still want me to come; but I shall understand, and admire, a message to the contrary.

Yours sincerely,

John Ziman

35

Solidarity within the Republic of Science*

It is the responsibility of the international scientific community to protect the Human Rights of its members, throughout the world.

Why are scientists being nagged so much about the fate of other scientists in other countries? Surely the 'outside world' of politics, war, and business has not become more wicked in the last few years? Yet there is a distinct feeling in the academic world that one can no longer decently reject appeals on behalf of this or that victim of oppression, in this or that country. The republic of science has never been ungenerous in its response to the needs of the scholarly refugee; but a new conscientiousness demands protests, boycotts, appeals and other uncomfortably positive actions on behalf of almost unknown individuals who have not yet even become refugees – who would, indeed, be very pleased if they could leave certain countries in health and sanity!

Every epoch of history has had its quota of oppression. In some regions of the world, tyranical régimes come and go, usually with the nastiest consequences for the unfortunate inhabitants but causing little stir elsewhere. Yet, in our own time, the conscience of scientists has been continually perturbed by the situation in one country where the bloodthirsty oppression of one of its worst tyrants has long ended, where there has been no change of government for more than a decade, and where, indeed, the scientific profession is relatively favoured. The fate of Soviet dissidents has become a very serious issue, not only in international politics but in the affairs of learned societies, in the columns of scientific journals, and in common rooms and laboratories.

The question is – should we do anything about it, and if so what? It is argued by many that scientists 'as such' should not be directly involved in these issues although as citizens they would loyally support the actions of their governments in these matters, and as private persons they might subscribe generously in money and effort on behalf of the victims. Others

* Published in *Minerva* 16, 4–19 (1978).

insist that the scientific community, national and international, cannot stand aside from events affecting the lives and careers of some of its members, and that this is a legitimate sphere of interest, even a serious moral responsibility, of the organized institutions of the scholarly world.

At first sight, this seems a very confused issue, where the arguments are evenly balanced. But the ethical principles underlying the debate have seldom been examined, and the historical elements in the situation have been completely ignored. Despite the absence of any dramatic shift of policy or circumstances, in the Soviet Union or in other countries, there has been a slow change of mood in the relationships between the scientists concerned and new perceptions both of realities and of responsibilities. In trying to decide how we ought to act, now and in the foreseeable future, we must look back several decades to observe these changes and to discount the influence of outdated simplifications and slogans.

Briefly, we find that we must redefine the traditional ethic of universalism in science, and reassert the social solidarity implicit in the concept of a 'republic of science'. This solidarity can no longer cohere around the simplified goal of 'the advancement of knowledge', but must recognize the significance of the social, political and legal conditions under which this goal is sought. To protect both the welfare of the individual scientist and the health of science itself, there must be a direct appeal to the international code of human rights, as a standard of justice, morality and corporate action.

In this light, the arguments for the involvement of scientists and scientific institutions in the human rights movement are overwhelming. This applies not only to the case of the Soviet Union and its associated régimes but with equal force to all countries – in Latin America, for example – where scientists are robbed of those rights and are suffering severe persecution. The cause of freedom and social responsibility in science demands a high degree of enlightened solidarity in the republic of science.

Interest in Russian science

The state of science in the Soviet Union is of more than passing interest to scientists in other countries. From the eighteenth century, Tsarist Russia was the home of a small but significant scientific community, linked to the major European centres of scientific culture. The other socialist countries of Eastern Europe share long and well-established traditions of science and scholarship with their neighbours to the West. Economic, industrial and educational developments since the Second World War have strongly favoured the quantitative growth of science in all these countries, so that no Western scientist can afford to ignore what may amount to 20–30 per cent of all published research work in his field.[1]

This interest goes beyond mere acquaintance with the relevant publications. The communication of information by direct spoken discourse – through conferences, lecture tours and exchange visits – is quite as important in science as written communication. This 'traffic' creates many occasions for personal contacts between individuals from both sides of the 'iron curtain', in cordial circumstances of cooperation and sociability.[2] Many Western scientists thus count Soviet scientists as personal friends, and have become quite familiar with the material and social conditions under which they work. The fact that this work is precisely similar to their own in its intellectual aims and instrumental techniques provides ample common ground for sympathy and understanding.

Curiosity was also aroused by the claim that research can be more effectively planned in the socialist system.[3] Partisanship and controversy in politics excite a genuine interest in the actual quality and style of science under these conditions, whichever side one may favour in the conflict of opinions. After the death of Stalin it seemed important and interesting to find out whether the facts were consistent with the shining propaganda surface behind which they were then almost completely hidden.

The cooperative ideal

Contacts between Soviet and Western scientists were re-established after the Second World War in an atmosphere polarized between two contrasting doctrines – on the one hand the universality and political neutrality of scientific knowledge, on the other the ideological conflict and technological competition of the 'cold war'.

The ideal of perfect transnationalism in science is deeply rooted, both in the actually operative norms which have been given formal expression by the sociologists of science,[4] and in many famous historical precedents. In the ethics of science much attention is focused on political and religious neutrality – the appeal to objective fact rather than the prejudiced opinions of sect and party in church and state.[5] These traditions are very real and positive, and can thus be invoked with the utmost sincerity by scientists on both sides of the 'iron curtain' to justify their 'cosmopolitan' tendencies and activities. At times this doctrine approaches a naïve belief in the supremacy of science over all other human values. Nevertheless, it is an enduring and fundamental principle, on which the very concept of a republic of science is based.

By contrast, crude nationalism and technocratic militarism are completely antipathetic to this view of science. If science were regarded as a major instrument of state power, then scientific contacts across the 'iron curtain' could only be judged subversive or reckless, and certainly not to be encouraged without very careful control by the appropriate governmental authorities. Thus, if there were to be any movement for such contacts within the scientific community, the principles of internationalism had to be continually emphasized, whilst reference to the contrary realities had to be completely suppressed. Indeed, one may say that it became a point of honour and of pride, in enlightened scientific circles on both sides, to favour international cooperation and peace with their scientific colleagues 'on the other side', even to the extent of repudiating the suspicions and antagonisms of their own governments. This was the spirit of personal solidarity within science that was evident, for example, in the Pugwash movement – although this was confined, in practice, to a relatively small group of scientific notables who were not altogether detached from governmental and other social responsibilities and loyalties.

The establishment of contacts proceeded, at first rather tentatively, through a succession of formal 'exchange agreements', and similar arrangements between governments or scientific academies or both. Individual scientists welcomed the opportunities for visits, which usually took place in an atmosphere of exaggerated cordiality. A number of major scientific congresses took place in the Soviet Union, and various projects for scientific cooperation were agreed upon. Indeed, it has become customary for a visit by a head of state or some other major political figure, or the signing of a treaty between the powers, to be celebrated by the announcement of projects of this kind; by a curious symbolic interpretation, reference to science is deemed to imply some commitment to peaceful transnationalism even between countries armed to the teeth with 'scientific' weapons!

For those who actually had to arrange scientific contacts across the 'iron curtain', the reality was not always so rosy. Governmental agencies unconcerned with science made difficulties about visas, foreign exchange facilities, freedom of internal travel, personal contacts and changes of programme, so that few such visits achieved anything like the sympathetic comradeship characteristic of international science in more open societies. The sheer inefficiency and complexity of some bureaucratic systems continued to baffle and fatigue those who tried wearily to get the machinery working. It became customary, indeed, amongst Western scientists, to make great allowances for the difficulties experienced by their Soviet colleagues in finding a way through their own system, and elementary discourtesies, such as failure to respond to letters, sudden changes of programme and late arrival, or unexplained absence, at conferences, were quickly forgiven. To complain about such a matter to a Russian friend, who obviously could not help it, seemed ungracious, whilst to make a public comment laid one open to abuse for 'trying to introduce an attitude of distrust to the possibility of Soviet–British scientific cooperation.'.[6]

Nevertheless, in the actual negotiations behind the scenes which established these agreements, a much more realistic attitude must have been apparent on both sides. Neither the Western nor Soviet scientific authorities, representing their national academies or governmental departments, could have been under any illusions about the type of bargain which they were getting, nor about the type of person with whom they had to make that bargain. The discussions which must have taken place have not been publicly documented but one can only suppose that superficial courtesy and friendliness covered a good deal of guarded toughness, which could not be melted by appeals to the universality of science or the solidarity of the scientific profession. And according to some of those who took part in such negotiations,[7] many cases affecting the human rights of individual scientists were introduced and favourably influenced by private persuasion during their meetings. In other words, those leading scientists and their administrative assistants actually responsible for scientific cooperation with the Soviet Union and the other socialist countries were always fully aware of the failure of science in those countries to live up to the ideals which were publicly professed on both sides, and they may even have shared, unconsciously, the cynicism which such hypocrisy tends to induce. In the circumstances of the 'thaw', this was almost certainly the most constructive policy which could have been adopted at that time, but its ethical dilemmas must not be glossed over if we are to understand the present situation.

Disillusionment

This tacit agreement not to comment on the state of science under communism began to crumble in the late 1960s. The official spokesmen continued to utter the same courtesies and compliments on official occasions, but the voices of disquiet were no longer muted. Scientists of standing began to add their names to public protests organized by groups which could not be regarded as especially hostile to communism or to the Soviet Union.[8] The leading organs of Western science, such as *Science* and *Nature,* now began to treat the politics of science in Russia and other communist countries with the same disrespect which they had always shown towards Washington and Whitehall.

This shift of editorial policy reflected a distinct change of attitude within the scientific community. During a period when the war in Vietnam was subjecting science to severe strains on ethical grounds, the mood of naïve scientistic good will towards the highly politicized science of the Soviet world could no longer be sustained. It was no longer so novel to visit Russia – and what was seen there was not always an inspiration towards closer cooperation. Exasperation at discourtesies, inefficiencies and occasional buffoonery could no longer be suppressed in the name of international peace and understanding.

Nor could one continue to gloss over the actualities of political repression involving science and scientists. For example, by the summer of 1968, Prague had become a favourite centre for international scientific conferences, several of which were brutally disrupted by the invasion of Czechoslovakia by Soviet forces. There was an immense wave of sympathy for the plight of the Czechoslovak scientific community, many of whose leading members were dismissed from their posts, driven into exile, or cut off from the international scientific network of communication and travel. The 'Prague Spring' and its aftermath had a peculiarly traumatic effect in the world of science.[9] No middle-aged European or American scientist can forget the course of the Second World War, nor the changing pattern of the map of Europe since 1935. He knows perfectly well that his contemporary from Eastern Europe has not lived all his life under communism, and is most appropriately treated as a Czech, a Pole, a Hungarian, or member of one of the other nations which share the European cultural and intellectual tradition. Bonds of sympathy and friendship have proved much easier to establish with the scientists of these countries than with Soviet scientists brought up in a far more unfamiliar culture.[10] Western scientists have seen tragedy strike amongst those whom they had come to value as intimate friends.

Soviet treatment of such distinguished scientists as Andrei Sakharov[11] and Benjamin Levich[12] contributed to the disillusionment. Many Western scientists who desperately wanted to avoid political issues could not ignore the 'stupidity' of the Russians in thus destroying their most valuable scientific assets. For what crime had such distinguished scholars been deprived of facilities for teaching and research? The realization that this was retribution for speaking with simple humanity, or for wanting to emigrate, was shocking. And it was not easy to put all the blame on 'external' agents, such as the secret police, when many genuine Soviet scientists – academicians, professors, research directors, etc. – had taken an active part in silencing their professional colleagues. Solidarity in the community of science could scarcely include those who might have betrayed all the norms of science itself by, for example, acquiescing in the removal of the published works of dissidents from the scientific literature.[13]

The Russian misuse of psychiatry for political repression[14] gradually seeped into the consciousness of the Western intellectual world, and finally aroused its conscience. But long before the decision of the World Psychiatric Association in Honolulu in August 1977 to censure it, this monstrous practice was affecting attitudes towards simple-minded cooperation. Both ethical and intellectual principles were evidently severely compromised in this cynical perversion of science and medicine.

Nevertheless, it is worth remarking that these crimes and follies are not really worse than those of previous periods. Hungary also suffered a Soviet invasion; many scientists of note were amongst the millions of victims of Stalin; Lysenkoism was as nonsensical, and had as devastating consequences both for science and for human beings in difficult circumstances, as Snezhnevsky's doctrines on schizophrenia. For some Western scientists who felt they were justified in overlooking these evils at the time, it has not been easy to change tack and to admit that they were mistaken. To some extent, the change of mood in the past decade can be attributed to the mere passage of time. The intense emotional devotion of a small number of leading scientists to the left-wing causes to which they had given themselves in the 1930s was respected by their colleagues, until a new generation found itself in responsible office throughout the scientific community.

Enlightened solidarity

This disillusionment with scientific cooperation could lead to a split, and withdrawal into isolation, with the world of science divided once more by an iron curtain. In an age of greater and greater technological interdependence of the countries of the world, this outcome seems unlikely; but it would be a tragedy for science if the transnational norm of universality were to be abandoned by either side in the name of moral purity or national interest. The ideal of solidarity within the republic of science has been damaged, but not superseded; it needs to be redefined in the light of our better understanding of political, legal, moral and scientific realities. On what grounds can some 'special relationship' between scientists in different countries be claimed?

Humanly speaking, friendships established through professional cooperation make strong demands on our sympathy. Scientists who read each other's papers, correspond about scientific questions, meet at conferences, and even undertake joint research projects, acquire mutual obligations which go far beyond the official protocols. In normal times, these obligations are readily discharged by free and open communication, by personal hospitality, and by all the other ways in which friendship is normally celebrated. But this relationship cannot then be quenched at the demand of a government official; to fail to act when this very colleague becomes an innocent victim of injustice is a betrayal of trust, an intolerable hypocrisy which would mock the concept of solidarity in its human dimension. In other words, Western scientists are more than ever concerned about the fate of their Russian colleagues whom they know as human beings beyond conventional civilities.

The scientific norm of universality also imposes severe obligations on all who would belong to the community of science. It is not a mere sentiment, but is a fundamental part of the process by which scientific knowledge is created and validated.[15] By its very nature, science is a cooperative activity, in which all who can contribute with observations, theories, concepts or criticism are free to participate. The transnational solidarity of science is thus strictly functional; Americans and Russians, Britons and Germans, Japanese, Brazilians and Nigerians, all belong to the same scientific community, and have a common interest in the welfare of all scientific institutions around the world. It is important that science should thrive in the Soviet Union and in other communist countries, for we rely both on the accuracy of the results of their research and on the acuity of their critical assessment of our theories. Once more, enlightened solidarity goes beyond proposals for joint research projects, and encompasses serious concern for the quality of scientific life in our respective countries. When we observe developments in another country which threaten the integrity of science itself, we are bound to speak out in the name of that solidarity. What is the alternative – to keep silent until science in that country has become so degraded intellectually that it can be ignored, and those who have to practise it need no longer be respected as colleagues?

Heightened awareness of fundamental political differences, of the contradictions of national rivalries, of injustices, political persecution, administrative follies and anti-scientific irrationalities does not, therefore, justify withdrawal into intellectual separation. On the contrary, interest in and knowledge about science in other countries is more necessary than ever. But because it poses ethical dilemmas, and tests our understanding of human, political and scientific realities, the practice of solidarity in the republic of science is not the simple game it seemed in the 1960s. The old scientistic slogans can no longer be trusted: we need a new ethical centre for our policies.

The human rights code

The slogan 'science is neutral' has great advantages for a learned society. 'Science', defined as the published results of research – thus avoiding the disturbing aspects of secret research and competitive technology – provides a ready-made category of non-political topics on which international cooperation may reasonably be sought. The enlargement of knowledge in these fields by conventional scientific methods is an uncontroversial goal without fundamental political implications. The long-term survival of science, scientists and scientific institutions in a hostile political or cultural environment has often seemed to depend on this prudential limitation of the agenda to relatively harmless technical activities related to rather academic themes. More immediately, there is no threat to the internal harmony of a national scientific community if there is an avoidance of policies in which controversial political issues come to the fore.[16]

For individuals and organizations to move towards a more enlightened conception of scientific solidarity, they need a framework of principles and programmes as coherent, universal and benevolent as those associated with the 'promotion of natural knowledge'. To enlarge the agenda of a learned society without reference to such a framework is to introduce terrible dangers of internal dissension and strife. Many men of good will have tried to persuade such bodies as the Royal Society to act conscientiously on behalf of scientists suffering oppression in many different countries: such actions have always been resisted by the argument that one would not then know where to stop along the slippery slope of political partisanship.

But humanitarian solidarity in science leads us to the framework of principles for which we are looking. When a British scientist says of a colleague in the Soviet Union that he has 'become an innocent victim of injustice' he is unlikely to be referring to the situation as judged by Soviet law. He may not even have in mind the specific provisions of British law. The injustice will almost certainly be seen as a violation of the human rights which have long been implicit in the constitutional legal order of most civilized societies.

Although the idea of human rights has a long history, it was first proclaimed as a common standard for all countries in 1948, when the United Nations unanimously adopted the Universal Declaration of Human Rights.[17] The essential elements of this declaration have since been embodied in a number of international conventions which have been ratified by a large proportion of the governments of the world. These conventions set up procedures for the consideration of complaints and for the enforcement of what is now a consistent international code. This code has thus become a universal standard by which we can judge the welfare of all men and women in all countries.

Many of the difficulties of achieving transnational solidarity in the world scientific community are thus overcome by appeal to the international code of human rights. This code is universal, it is phrased in precise legal language, and it has been accepted in principle by most civilized governments. Actions based on this code thus stand above political squabbling and the conflict of governments. A learned society which takes up the cause of foreign scientists whose human rights have been infringed can scarcely be accused of partisan political action; on the contrary, failure to act in such cases could be regarded as neglect of a moral duty.

Opposition to action by scientists on behalf of foreign scientists is often based upon the objection that scientists as such should not claim special privileges above those of other citizens. Why should a theoretical physicist, for example, expect stronger support for his plea for permission to emigrate from the Soviet Union than a bus driver? Under the articles of the code of human rights, these cases are of equal merit. A group of British physicists might wish to exercise pressure on behalf of their Russian colleague because they feel bound to him by ties of friendship and professional solidarity; but no additional right is demanded for the scientist which would not be justifiable for the bus driver in similar circumstances. The code of human rights provides precisely that universal framework of principles within which solidarity can be realized in practice without any suspicion of particularistic interest or of a 'cosmopolitan conspiracy'.

Human rights and scholarly freedom

The international code of human rights is primarily intended to protect the individual against various forms of exploitation and tyranny. But many of its provisions go much further than guaranteeing 'life, liberty, and the pursuit of happiness'; they enunciate the conditions which are necessary for science itself.

Enlightened solidarity within the republic of science implies concern about attacks on the fundamental practices of the scientific enterprise – practices such as freedom to teach and to publish the results of research, freedom to criticize the research of other scientists, rights of communication and assembly, opportunities to visit foreign scientific institutions and to meet foreign scientists, respect for priority of discovery and for established scientific authority, access to education without discrimination, and opportunities for other forms of participation in the scholarly enterprise. These practices are not claimed to be unique to science and other branches of learning. In fact, they are no more than particular forms of the essential conditions for an open society – and as such they are protected as human rights.

It is a remarkable fact that all the needs of scholarly freedom are explicitly covered by the universal code of human rights.[18] This coverage is almost perfect. A complete list of the legal and social conditions required for the pursuit of science can be found in the articles which have been thoroughly discussed, debated, legally defined and diplomatically negotiated into one or other of the conventions on human rights.

Thus, for example, what more is needed to protect the traditional practices of scientific communication than Article 19 of the Universal Declaration of Human Rights: 'Everyone has the right to freedom of opinion and expression; this right includes freedom to hold opinions without interference, and to seek, receive and impart information and ideas through any media and regardless of frontiers.'

The corresponding articles of the later conventions elaborate, in legal phraseology, the acceptable limitations to this right: for example, in the interests of national security, for the prevention of disorder or crime, and for the protection of the reputation or rights of others. But these restrictions do not weaken the applicability of the principle, for they define the extent to which it may be subordinated to other principles with which it may in practice conflict.

It is realistically argued that these conventions, like the corresponding provisions in the constitutions of many totalitarian countries, are mere window dressing, without practical effect. In many cases, there is little prospect of actually enforcing them through the machinery of international law. But the fact that these articles have been publicly accepted by the governments in question is always a valuable factor in a campaign of protest against their violation.

The universal code of human rights thus defines an unimpeachable standard of behaviour to which all governments concerned with the advancement of science and learning may be expected to conform. If they were to behave according to this code, then the pursuit of science would be adequately safeguarded. Alternative attempts to define the practices, conventions and rights necessary for a healthy scientific community are scarcely likely to improve on those already provided for in the international code of human rights, which is now legally recognized by almost all civilized countries.

The obligation to act

The human rights code is not just a distant ideal. From experience in many democratic countries, we know that it is capable of substantial realization – imperfectly, partially, but in the genuine spirit and in many particular details. To give effect to the code is thus much more than an aspiration: it has become a moral imperative, as emphatic as the obligation of every citizen to uphold the laws of his own country. This applies with particular force to scientists and scientific institutions, who belong as much to the world community of science as to their own societies.

Until recently, the difficulty of redefining the limits of solidarity in science has inhibited its extension beyond conventional 'cooperation'. With human rights as a criterion, enlightened solidarity within the republic of science can now be given form and substance. The inconsistencies, contradictions, soul-searchings and hypocrisies associated with the scientistic policy of 'all science is good' can now be faced. It is, indeed, the duty of scientists and learned societies to affirm and practise the true principles of transnationalism for the good of humanity and the advancement of knowledge. In times of intense national rivalry and conflict, this responsibility is graver than ever, independently of talk about 'détente' and other catch-phrases in the diplomatic sphere. But if the scientist is not to be judged inhumane and concerned only about his own well-being, then the principles to be upheld must be those of the rights of man, which protect him as an individual and as a member of the profession of science itself.

During the past decade, many learned societies have felt pressures from their members to act openly on behalf of scientists suffering oppression. This pressure has often been resisted in the name of 'the unity of science' and 'political realism'. These slogans are now so discredited that new principles of policy will have to be found. Any scientist who has begun to understand the true meaning of solidarity in the world scientific community will find it quite distasteful to take any further part in the empty ceremonies and insincere compliments which have often accompanied 'scientific cooperation' in recent decades. A shift to the principle of active enlightened solidarity based on the code of human rights is not merely a moral duty; it will soon be seen as expedient by the 'realists' who have to govern the institutions of the scientific community. That is the only way that they will be able to satisfy the moral conscientiousness and scrupulousness of their members, and give them leadership amidst doubt, shame, anger and conflict.

What could be done?

The most important step towards action on behalf of the victims of oppression is to accept that failure to act would be a moral dereliction. The means and the forms of action then follow. Individuals and informal organizations have already shown, by protests and demonstrations on a pitifully small scale, that much can be achieved even by the most modest efforts. Imagination and experience have taught the power of various techniques of protest, in various circumstances, against various targets. This is an expertise acquired by practice, and by informal observation, rather than from any general theory.

But no amount of agitation by individual scientists or small groups can match the pressure which can be applied by an established learned society. How should that pressure be applied? For a long time it has been argued by the officers of such societies that 'private persuasion' is the most effective policy.[19] In the era of 'cooperation', this policy of trying to influence the fate of individual victims of persecution by informal discussions with the officers of national academies or governments, during the course of celebrations or negotiations, did perhaps have its successes. It will still have its place under the more strenuous conditions of enlightened solidarity.

But private persuasion can only be really effective if it is backed by a strong public position, affirming beyond doubt those principles which cannot be compromised. This affirmation is essential, not only as a signal to those suffering oppression that they are not being forgotten, but also as a wall at one's back in any secret negotiations. A deficiency of past policies of official Western science in relation to Russian science has been unwillingness to make such public statements of principle, for fear of charges of 'provocation' or 'non-cooperation'.[20]

Enlightened solidarity, however, demands more positive action. It requires activity in collecting information about policies, and their victims. Nothing has been more dispiriting to those deeply concerned about the state of science, and the fates of scientists around the world, than the apparent absence of the slightest flicker of interest in these matters in the highest scientific circles. One might have thought that the officers of the great learned societies would have informed themselves fully on all aspects of scholarly affairs in other countries, and would welcome further information and opinions from any reliable sources. Current changes in this attitude are welcome evidence that scholarly freedom and human rights are indeed the concern of all scientists, and convey that message both to oppressed and oppressors.

But the great question is the form and possible extent of formal protests and sanctions. This question cannot be answered in the abstract; enlightened solidarity is an ethical principle, not a plan of campaign. All that we can say is that this principle can fully justify statements of condemnation, and acts of 'non-cooperation' against governments, scientific institutions or individual scientists who seriously violate the code of human rights in ways which threaten the advancement of knowledge and the pursuit of science. Without willingness to go to such lengths, if necessary, in support of this principle, all assertions of good will and sympathy pay no more than lip-service to the needs of the oppressed.

Needless to say, any such decision by a responsible individual or learned society must be taken cautiously and deliberately. The action must be appropriate, efficient, and proportionate to the evil to be combated.[18] The exercise of enlightened solidarity is not simple and automatic; practical experience and wise leadership are needed to decide how far to go in each case. It is a very serious question, for example, whether to break off relations with the administrative organs of a national scientific community on behalf of a small number of oppressed scientists in that country.

An important factor, also, must be a properly balanced policy which does not seem to be directed specifically against any particular ideological or political system. At the present time, for example, violations of the human rights of scientists in certain Latin American countries under military dictatorships are quite as shocking as those which take place in many countries under Soviet domination, and demand equal condemnation, and equally determined action. That is not to say that the same action is appropriate against a violent and bloodthirsty dictatorship, ruthlessly suppressing its supposed opponents in the wake of a *coup d'état,* as against the soul-destroying pressures of a totalitarian bureaucracy corrupted by long years in power. Account must be taken of the difference between a very weak scientific community trying to establish itself in a less developed country, and the immense, highly organized, and extremely able scientific establishment of an industrialized state such as the Soviet Union. These are matters which would have to be taken carefully into consideration in any decision to protest, or to take sanctions in the name of scientific solidarity.

The need for reliable information

In all discussions about what should be done about the victims of oppression, the same question arises: are we sure of the facts? In the absence of proper courts of law which can make inquiries over national frontiers, this objection to any action can always be raised. By the same token, the oppressors themselves are in a strong position to deny that any injustice has been done, and to spread such lies as they can make credible at home and abroad.

Some of the distaste for any concern with issues of human rights stems from these unpleasant conflicts of assertion and evidence. For a humble scholar, who is brought up to believe that he must only act on the soundest knowledge of the facts, these unseemly and emotive controversies are especially repellent. It always seems safer to defer any positive action until 'the real facts are known' than to step into all the mess and muddle of law and politics. Such caution is very often a mere rationalization of unwillingness to accept a moral responsibility, but it must be met as far as possible.

An important aspect of scientific solidarity is, therefore, the collection and scrupulous scrutiny of all available evidence on all cases of persecution of science and scientists.[18] A great deal of evidence is, in fact, in the hands of a variety of persons and organizations with immediate personal concerns about the fate of their countrymen, or is accidentally revealed to visitors and correspondents. Thus, the main facts about the Soviet abuse of psychiatry were already ascertainable, beyond reasonable doubt, by the time of the meeting of the World Psychiatric Association in Mexico in 1973.[21] The obstacle to concerted public action, in this and other cases, was that this information had not been collected, collated, sifted, and validated by any independent body whose legal and political credentials were beyond question. It is not sufficient, in issues of such weight, where promptness of action is quite as important as sheer force, to rely upon interested parties to bring up the issues and eventually to get some action within the ordinary framework of the constitution of a learned society. In many learned societies, even the machinery for initiating debate on such a matter in a governing council or at a meeting of members is very slow and cumbersome, and can often be stalled by determined internal opposition.

There is an urgent need, therefore, for a clearing-house for information about the violation of human rights in the world scientific community. This body needs both scientific and legal expertise, and many channels of communication, open and secret, with individuals and organizations, inside and outside the scientific world. Its job would be, simply, to collect relevant information, and to evaluate it in the light of the code of human rights; from it would come reliable accounts of individual cases on which scientists and their organizations could justifiably take action.

Steps are already being taken by various bodies to collect and distribute such information.[22] It is incumbent upon the republic of science, in the interests of its own solidarity, to support such a clearing-house at the transnational level, with the strongest possible facilities, authority and integrity. This responsibility falls directly on the International Council of Scientific Unions, which is the worldwide organization in which all learned societies and national academies of science participate. But because of the legal aspects of the code of human rights, this responsibility needs to be shared with the International Commission of Jurists, which has wide experience and international respect in the impartial assessment and evaluation of human rights issues. An authoritative organization of this kind, under the aegis of the International Council of Scientific Unions and the International Commission of Jurists would be an invaluable instrument for a genuine policy of enlightened solidarity in science.

Conclusion

This discussion seems to have moved a long way out of the historical, political and geographical context of the relationships between the scientific communities of the communist and non-communist countries of Europe. It has entered into the ethical domain, with very little reference to the very serious particular problems of present-day corruptions and evils.

But experience over the past decade has shown clearly that all attempts to react to these particular problems have been hamstrung by the lack of an agreed body of basic principles and ideals. Confusions and contradictions in well-intentioned proposals have tangled the skeins of motive and policy, so that little is achieved – to the advantage of the oppressors and their friends. It is only by a return to ethical foundations that the issues can be clarified and a consensus on policy created. It is only by reference to these foundations that we can act with justice on behalf of Soviet scientists and Soviet science.

Even now, patience will be needed to develop a policy of enlightened solidarity within the republic of science, and to make it work. It is not a perfectly simple policy, to be explained in a few words and grasped in a moment. It has subtleties and ambivalences which cannot be decided by slogans or by reference to an encyclopaedia. But it is based upon our common humanity, and upon the social relations which have generated scientific knowledge, which is amongst the greatest of human achievements.

References

1. By contrast, most scientists are profoundly indifferent to the state of science in poor countries such as Venezuela, Nigeria or Indonesia, whose scholarly problems are very grave, but whose contribution to knowledge has so far been very small. See Moravcsik, M. J. and Ziman, J. M., 'Paradisia and Dominatia: Science and the Developing World', *Foreign Affairs* 53, (4) 699–724 (1975).
2. This is not to deny the serious deficiencies of practical participation by the Soviet Union in these activities over the years, even since the 'thaw' of the 1960s. See Ziman, J. M., 'Letter to an Imaginary Soviet Scientist', *Nature* 217, 123 (1968). These practices have not changed: see Nye, J., 'Russians at Conferences', *Nature* 249, 8 (1974).
3. This was claimed, for example, by J. D. Bernal whose book *The Social Function of Science* (London: Routledge, 1939) anticipated both the study of 'science policy' and the modern movement for 'social responsibility in science'.
4. Merton, R. K., *The Sociology of Science* (University of Chicago Press, 1973), pp. 267–78.
5. This was always the deliberate policy of the Royal Society of London from its foundation in 1663, though it is nowhere specifically laid down in its charter of statutes.
6. Engelhardt, W. A., 'Letter in reply to "Letter to an Imaginary Soviet Scientist" ', *Nature* 218, 404 (1968).
7. For example, Professor Sir Harold Thompson, former foreign secretary of the Royal Society, on numerous occasions, public and private.
8. For example, the special issue on 'Détente and Soviet Science', *Newsletter of the Federation of American Scientists* (October 1973).
9. See Janouch, F., 'Science under Siege', *Bulletin of the Atomic Scientists* 32, 6 (1976).
10. Yet there has been a slight tendency to ignore the plight of science in some of the smaller Eastern European countries, relative to the more dramatic issues and larger stage of dissidence in Russia itself.
11. The first reference to Sakharov as a dissident which I can find in the specialist scientific press is Campbell, L. 'Sakharov: Soviet Physicist Appeals for Bold Initiatives', *Science* 161, 556 (1968). The famous cable from the National Academy of Sciences of the United States to its opposite number in the Soviet Union expressing

'deep concern' about the treatment of Sakharov is reported in *Science* **181,** 1148 (1973).

12. Shapley, D., 'Soviet Science: Levich's Delayed Emigration Stirs Concern', *Science* 176, 1314 (1972).
13. See, for example, 'USSR', *Nature* 268, 95 (1977).
14. See, Reddaway, P. and Bloch, S., in *Russia's Political Hospitals: The Abuse of Psychiatry in the Soviet Union* (London: Gollancz, 1977).
15. See, Ziman, J. M., *Public Knowledge* (Cambridge University Press, 1968).
16. A typical current issue is whether American criticism of Soviet treatment of dissenters would be damaging to détente: see, for example, Walsh, J., 'Soviet–American Science Accord: Could Dissent Deter Détente?', *Science* 180, 40 (1973). There is an implicit acceptance of a non-scientific political criterion – that is, that 'détente' is a desirable objective in itself – by both sides when this issue is debated within the scientific community. In other words, scientific cooperation is being justified for its contribution to a particular type of political relationship between the two countries, not for what it can do for politically neutral science.
17. For the full text of this and other instruments of the code of human rights, see Brownlie, I. (ed.), *Basic Documents on Human Rights* (Oxford: Clarendon Press, 1971).
18. See Council for Science and Society, in collaboration with the British Institute of Human Rights, *Scholarly Freedom and Human Rights* (Chichester: Barry Rose, 1977); reprinted in Reports and Documents, *Minerva* 16, 283–326 (1978).
19. See, for example, Norman, C., '. . . in Public or Private?', *Nature* 259, 167 (1976).
20. This deficiency had to some extent been remedied recently by the resolution by members of the United States National Academy of Sciences on freedom of inquiry and expression on 27 April, 1976, and by the presidential address of Lord Todd, P.R.S., at the anniversary meeting of the Royal Society on 30 November, 1976.
21. Reddaway, P. and Bloch, S., *op. cit.*
22. For example, by the Committee for Scientific Freedom and Responsibility of the American Association for the Advancement of Science.

PART SEVEN

THE COMMUNICATION SYSTEM OF SCIENCE

36

Information, Communication, Knowledge*

Current deficiencies of the communication system of basic science cannot be made good by computerization; the need is for greater human effort in the production of review articles and monographs.

Of all the trendy, cliché words, 'communicate' is about the squarest. When E. M. Forster invented the phrase 'only connect' as the dedicating 'quotation' for one of his novels, he was talking of the human heart, not about the spewed-forth products of the head-office duplicator, nor the glossy gaudiness of the Sunday supplement.

But, for scientific research, communication is essential. I will not spell out here my contention that science is to be characterized as 'public knowledge'; in other words, that the aim of the scientist is to create, criticize, or contribute to a rational consensus of ideas and information.[1] If you accept this as a general notion, you will agree that the results of research only become completely scientific when they are published. All practising scientists know this, though they are often more careless with their grammar than they would dream of being with their galvanometers.

It keeps being said, generation after generation, that the then current system of scientific communication is in a dreadful mess, and that something ought to be done about it. Century after century, nothing much does get done about it, except that it swells to ever greater bulk.

There are two simple interpretations of this state of affairs. Those of us who are radicals at heart will blame the innate conservatism of humanity, and say that we continue to put up with a grossly inefficient and cumbersome system because we are too lazy to make a change, or too unimaginative to appreciate the tremendous advantages that a new system would offer. The natural conservatives, on the other hand, will deny that the system is as bad as all that, and point to the good sense of the scientific community in not having been stampeded into some extravagant and newfangled system whose untried benefits exist only on paper.

* Address to Section X of the British Association, Exeter meeting, 1969; published in *Nature* 224, 318–24 (1969).

As a contribution to this debate, I shall attempt first to justify the present system – that is, to set out explicitly its unconscious rationale. A social institution that has existed for several centuries must have an intelligible function; when we understand that properly, we can ask how the present practice ought to be modified to continue to achieve the desired goals.

Our present system of scientific communication depends almost entirely on the 'primary' literature. This has three basic characteristics: it is fragmentary, derivative and edited. These characteristics are, however, quite essential.

The invention of a mechanism for the systematic publication of fragments of scientific work may well have been the key event in the history of modern science. A regular journal carries from one research worker to another the various discoveries, deductions, speculations and observations which are of common interest. Although the best and most famous scientific discoveries seem to open whole new windows of the mind, a typical scientific paper has never pretended to be more than another little piece in a larger jig-saw – not significant in itself but as an element in a grander scheme. This technique, of soliciting many modest contributions to the vast store of human knowledge, has been the secret of Western science since the seventeenth century, for it achieves a corporate, collective power that is far greater than any one individual can exert.

Scientific papers are derivative, and very largely unoriginal, because they lean heavily on previous research. The evidence for this is plain to see, in the long list of citations that must always be published with every new contribution. These citations not only vouch for the authority and relevance of the statements that they are called upon to support; they embed the whole work in a context of previous achievements and current aspirations. It is very rare to find a reputable paper that contains no references to other research. Indeed, one relies on the citations to show its place in the whole scientific structure, just as one relies on a man's kinship affiliations to show his place in his tribe. All this becomes perfectly natural, and proper, once one has accepted that normal science is a highly cooperative activity, the corporate product of a vast social institution, rather than a series of individual forays into the unknown.

The editing – in fact, the censorship – of the scientific literature is a more delicate matter. It occurs in two stages. In the first place, the author presents an entirely false picture of his actual procedure of discovery. All the false starts, the mistakes, the unnecessary complications, the difficulties and hesitations, are hidden; and a yarn, of preternatural prescience, precision and profit, is spun. By the vision of hindsight, all is made easy, simple – and apparently inevitable. A scientific paper is not a candid autobiography, but a cunningly contrived piece of rhetoric. It has only one purpose; it must persuade the reader of the veracity of the observer, his disinterestedness, his logical infallibility and the complete necessity of his conclusions.

External censorship of scientific papers is, of course, the function of the anonymous referees. As you will learn at any laboratory coffee club, the editing and refereeing of journals is a fertile source of folk lore, anecdotes, grumbling and bad feeling.

I would argue, nevertheless, that this also is an essential element of our system of scientific publication. We must be able to rely on the basic accuracy and honesty of what we read in other people's papers, for we are always using their results in the construction of our own researches, and simply cannot find the time to repeat all their experiments, measurements, calculations, or arguments for ourselves. Scientific authors are often wrong, but their errors are usually made in good faith. I cannot see how this innocence could be preserved, against careerist pressures to publish, if there were no scrutiny by expert referees. The communication problem would be ten times worse if we had to wade through tomes of irresponsible nonsense in a search for a few reputable papers. Even the most commercial of daily newspapers distinguishes in its layout between genuine news and mere advertising. It is true that an experienced scientist 'internalizes' the critical standards of the referees, and can be relied on to be honest, accurate and reasonably plausible, in all his writings; but that does not justify the abolition of the referees – any more than the habitual honesty of middle-aged bank clerks would justify the abolition of the fraud squad.

These are the characteristics of the primary system of scientific communication. What are the defects?

To some extent, they are contradictory. One demand is for greater speed of publication. It is argued that the strategy of the individual research worker is so strongly influenced by the information that he receives from his contemporaries and rivals, that a good deal of scientific effort is wasted by multiple discovery or by one scientist being unaware of relevant results that have already been obtained elsewhere because of the delays of publication.

The supposed wastage from this cause is somewhat exaggerated. A simultaneous discovery by independent researchers is not necessarily such a bad thing; it is disappointing for those concerned, but it is an admirable form of confirmation of the supposed result. It is seldom that there is just a particular discovery to be made – like finding the very sixpence that one has just dropped – but rather a complex of ideas and phenomena which may be illuminated in different ways by different investigators. Nobody, surely, regrets that Heisenberg and Schrödinger should have simultaneously 'discovered' what turned out to be the same quantum mechanics; they arrived by somewhat different routes, and alighted at different stations on the way, so they have somewhat different stories to tell us about their journey. Obsession with sheer immediacy and speed is a neurotic symptom, mainly to be observed in certain fashionable fields where too many people are chasing too few ideas.

I am talking here about 'pure' science without an immediate application. The situation is rather different in applied research and development, where the best available solution to some particular problem is being sought, and where the rewards of success – or the cost of failure – may be measured in millions of pounds. The hindrances to the immediate spread of new knowledge are not then so much in the machinery of publication, but rather the barriers of secrecy deliberately raised around industrial and military research. The communication system of technology is quite different from that of pure science, having different ends, different norms and altogether different standards of morality. Applied scientists and technologists do, indeed, seek a great deal of information from the literature of basic science, but my impression is that they are seldom sufficiently instructed to be able to benefit from the very latest, as yet unpublished, research findings.

What are the natural time constants of research? In general, I would confirm the observations of Garvey,[2] who found that the various stages of hypothesis, design of apparatus, experiment, testing, confirmation, critical analysis, informal discussion, writing up, and so on, take months or years to complete, so that the interval of about four months between the receipt of a typescript and its publication in a reputable journal is not a significant proportion of the time required to 'make a discovery'. Only about one paper in ten thousand is so startling as to set the scientific world a-jangling, so that a very small number of quick-publication journals would be quite enough – provided that all the little shepherd boys can be taught not to cry 'Wolf!' too often.

Another development, which has been accorded almost official approval,[3] is the systematic exchange of 'preprints' of papers that are being submitted for publication. This procedure has two roots; it stems from the ancient courtesy of writing private letters to distant colleagues in advance of the official journals, and from the custom, developed during the Second World War, of circulating reports (often secret) of the results of researches to all interested parties, quite independently of the usual channels of publication.

In itself, there is no great harm in this custom, which can often serve a useful purpose. What was wrong was the proposal to develop 'preprint exchanges' that would attempt to make this haphazard custom truly efficient. Each author was to send his typescript to a central office where it would be duplicated and sent out to everybody else in the group. Membership of the group would be controlled by some procedure of authentication and introduction; no doubt after the sponsoring body had grown tired of subsidizing the duplicating and office costs, subscriptions would have been solicited from the various contributors. It is clear that a marvellous discovery had been made by the proponents of this scheme: they had invented, not the wheel, but the scientific journal of a learned society. Given another decade or so, they might even have rediscovered the referee!

To my mind, it is tragic that so many excellent and experienced scientists should have fallen into this trap, and encouraged vast expenditure on preprints and such like, instead of putting all their weight into the reform of the existing learned societies and the improvement of their journals. As I have said, the necessary delays occasioned by refereeing and decent printing are not really gross; a journal that gets a year behind in publication just needs a thorough kicking by its subscribers and contributors; to supersede it by yet another machine liable to just the same faults is sheer folly.

At the heart of the matter is the fallacy of giving too much significance to the role of 'informal' communication. Of course this is very important indeed – all those private letters, conversations over drinks, chatty seminars, conferences, meetings, lecture tours, and such like. It has always been understood that much scientific knowledge actually travels in this way – especially those key attitudes, techniques and insights that can only be transferred by prolonged contagion.

But the informal cannot, by definition, be made specific and rigid. One of the major purposes of the whole scientific enterprise is to draw from the confused, vague, inchoate 'stuff of experience' a few precise, clearly defined, 'objective' (that is, if I may use the term, 'consensible') concepts, principles, or observations. It is essential that scientific work should be 'written up' in full, with all the details of technique, interpretation and logical limitation necessary to persuade the reader of the truth of the conclusions – or at least sufficient for him to repeat the experiment or calculation for himself. The primary scientific literature is sometimes referred to, slightly deprecatingly, as 'archival', as if it were a sort of antiquarian record of deeds done and now finished. That is very far from the case. There is a long period during which they are active documents, cited by other workers and relied on in detail for later developments. A paper is useless unless it has these qualities of precision, for it cannot otherwise be accorded the critical judgement of the scholarly community, and thus be accepted or rejected as part of the whole body of scientific knowledge. I am not denying the value of hints, intuitions, speculations and other means of spreading new ideas; but it must always be clear, in the mind of the listener or reader, whether or not, so to speak, the witness is on oath. Any reputable scientist who allows his research work to be widely published in permanent form must take the consequences for any errors that it contains; he cannot repudiate the responsibility for misleading his colleagues by saying that of course it was only an 'informal' communication, and hence to be understood only in a Pickwickian sense.

The other great objection to present procedures of scientific publication is the vast bulk of material that is printed and has to be stored on the library shelves. More serious than the cost of infinite numbers of new bookcases is the difficulty of finding out what each of these papers has to say, and deciding whether we ought to study it more closely. This is called the information explosion, and the problem is said to be that of information retrieval.

The conventional system is not without its resources against this suffocating foe. The standard weapon is the abstract journal, where short summaries of all the papers in the current primary journals are printed in classified order, so that they can be noticed when they first appear, or rediscovered through the index if they happen to be forgotten. The first charge on the foundation of a new scientific laboratory should be a subscription to the appropriate abstract periodical, which will need to be consulted daily by the scientific staff. This sort of publication is indispensable to the scholarly world, for it is the only overall index to the archives of the discipline.

The tasks of abstracting, translating, classifying and editing such publications demand a high degree of skill and scientific knowledge. The cost of such organizational skill is not negligible; there has been a tendency to take for granted that it can be done on a shoestring, in spite of the tremendous increase in the bulk and complexity of the material to be dealt with. Most scientific papers now carry abstracts prepared by their authors, very often with translations into the major scientific languages, so that some of the burdens have been eased, but it is still a big job that has to be done to the utmost of perfection.

It is not surprising that such systems have attracted the attention of the technicians of data storage – the computer engineers. 'Why,' they ask, 'should one bother to have all this stuff printed in great fat books, when it could be stored so economically on magnetic tapes, and called up for instant perusal by pressing the appropriate buttons? Here is an activity ripe for electronic mechanization.'

The first obstacle is that the amount of 'information' to be stored is very large and miscellaneous. What the searcher wants, eventually, is a chance to glance at the text of every scientific article relevant to his particular problem. Nobody is going to go to the tremendous expense of putting onto magnetic tape the full text, or even the abstracts, of all the scientific journals that have ever been printed. The labour of setting them into type in the first place was expensive enough; who is going to press all the teletype keys those billions of times so that the words may eventually be reproduced once more, rather less elegantly, as computer print-out? The fact is that in a good scientific library, with the journals well bound and properly classified, it is remarkably easy to collect together the actual volumes one needs, and lay them open on the table for this sort of perusal. The mechanical storage of documents by microfilming, video-tape, or other means cannot yet compete with the shelf of books for occasional access – if you know what you are looking for!

The emphasis then shifts to the indexing and classifying phases of the operation. Can a machine system be devised that will turn up the addresses of all scientific articles that are relevant to the problem in hand? This is an extremely interesting question. It turns out to be only part of the general problem of the classification of information. A computer can certainly be used to search for complex combinations of topics described by a collection of key words – for example, 'electron', 'localized', 'disordered system' – which could not easily be located with precision in an ordinary classified index. There are tricks that can be played, like going through a list of approximate synonyms of the topic names presented by the inquirer. It is even possible (though expensive and laborious) to have the computer scan through the texts of all the abstracts in its store, so as to throw up, say, all the articles that contain the word 'percolation'.

But all such procedures are subject to the fundamental law of information retrieval – as the recall ratio increases, the precision ratio decreases. In other words, if you try to get hold of nearly every paper that there is on your chosen subject, you will be faced with a collection in which a high proportion of the material is obviously irrelevant; in your attempt to find out all about the percolation of electrons in semiconductors, you will be inundated with papers about percolation of water through the foundations of dams! As with the analogous problem of machine translation, the difficulties arise from the imprecision and metaphorical flexibility of language. The limitations of a particular system of automated retrieval will really depend on the nature of the scientific information being sought, and (as with the most elementary indexing operation) the fundamental skills of those who put the data on the tapes in the first place.

It is thus by no means certain that automatic information retrieval systems should immediately replace the conventional abstract journals with their comprehensive classified indexes. Computerized typesetting is playing an important part in the simultaneous compilation of primary journals, indexes, abstract journals, 'title' journals and their various author and subject indexes, but this is only a technical refinement of the conventional system. There are also very interesting possibilities for the retrieval of precise data in those branches of science where exact description and classification are rigorously prescribed, such as biological taxonomy and structural chemistry.

Another use to which computers have been put is in the preparation of citation indexes. This is a new device where references are given for all papers that have cited a particular paper published previously. By revealing the interconnexions of the scientific literature so clearly, this sort of index is of great interest to those who study the scholarly community at work, and it can also be used to search for possible relevant material, as a supplement to the ordinary abstracts. Unfortunately, it is a very expensive publication, which is not yet worth its marginal utility except to the very largest institutions.

The real reason why computer storage is not very efficient in these circumstances is that verbal information is a rather passive material, which can only be processed effectively by the human mind. There is so very little that one can really do to a classified index of abstracts on tape, except to run through it now and then looking for particular items. The specific technical advantages of a computer – the power to perform vast numbers of such operations at enormous speed – are seldom exploited. We can ask it to produce six different kinds of index – by name of author, by subject, by institution, by date, by journal, by country – and then what? Some of the writing on automatic information retrieval is really an attempt to invent more uses for it to justify the enormous initial cost.

Much effort, for example, has been expended on the idea that what everybody really has been waiting for is a daily alerting service that will tell each one of us about the latest papers in the topics which interest us. Thus, I shall have given instructions, on a punch card, to be kept up to date on the 'theory of liquids', on 'electron theory of metals', on the 'structure of alloys' and on 'lattice conduction in solids'; hey presto, the latest abstracts or papers on just these topics will be waiting on my desk each morning by courtesy of Electronic Alerting Services for Yokels (EASY). But I don't really care that much about what everybody is doing in all these subjects; and anyway I have not the time to read through all this stuff carefully; and also I've gotten interested in spin waves again recently and have forgotten to inform EASY about this; and what I suddenly discovered last week was a paper published ten years ago which gave the solution to my problem in another form; and anyway I'm damned if I'm going to become the slave of this sort of guff.

The conventional system has another weapon up its sleeve: specialization. The bulk of science grows by a complex process in which old journals expand, and then subdivide, and new ones also arise in the interdisciplinary regions. It may even be claimed that the modern scientist need not read any more journals than his predecessors – he simply narrows his vision until he takes in about as much material as before, over a more limited and specialized range. A journal that begins to specialize in a particular topic attracts many papers on that topic, and is soon able to set a high standard that reinforces its prestige.

Indeed, this can have most serious effects on the development of good scientific standards in many countries. The local journal of the national academy has too small a circulation, and is too diffuse in subject matter, to be an attractive vehicle of publication for the scientists of that country. They in their turn, by insisting on publishing their work in foreign or vaguely international journals, rob the local journal of its best material, thus depressing the apparent standard of research in that country.

Or is it now too late to save the general journals of the national academies and learned societies? Is it not more realistic to recognize that specialization of the organs of primary scientific publication serves a real purpose and should be encouraged rather than resisted?

Of course, our friends the root-and-branch boys have not failed to put forward a 'rationalized' scheme as the logical conclusion of this line of development. They would have us create a single centralized bureau of scientific publication, to which would be despatched typescripts (or would it be transcripts of telephone messages?) to be edited, refereed, classified, printed and sent out in exactly preselected categories to licensed subscribers. I do not think I need to set out the objections to this modest proposal; it would be monopolistic, bureaucratic, fundamentally inefficient, blinkering and inflexible; scientific knowledge is not like bread, that can be manufactured by machines and distributed on a very large scale, provided that no-one minds if it is as tasteless and formless as cotton wool; food for the mind requires to be prepared with a human touch.

Returning to the problem of finding one's way through the vast jungles of the literature, a continuing puzzle and offence to sensible scholars is the vogue for the publication of the papers presented at scientific conferences. One can understand the vanity of the organizers and participants, who love to see themselves in print, and the cost–benefit accountancy of administrators who feel they must have some tangible return for the expenses that have been incurred in getting everybody together for a few days – but why should anyone be willing to buy an expensive volume of short communications, inferior versions of proper papers, published several years after they were written and delivered? The answer is simple: a well planned conference, being a meeting of most of the 'invisible college' of experts in a particular topic, is an occasion for general stocktaking, and the proceedings can be expected to give a fairly complete coverage of results achieved and work in progress at that epoch. In other words, a conference report, although inferior to conventional publications in speed, quality and accessibility, can be used very effectively as a starting point for further search or research. Conceived apparently as a medium for quick, informal communication, it becomes a means of attack on the bulkiness of the literature.

But surely this need is much better met by review articles deliberately commissioned for this purpose? The growth of this class of secondary literature is an important feature of the conventional system of scientific communication in the past half century – a part of the machinery that is often ignored by the technocrats, but which really needs to be greatly strengthened.

In its narrow sense, a review article is little more than a classified bibliography – a *catalogue raisonné* of the primary literature, putting the results in order and commenting impartially on any obvious contradictions or controversies. The author is thought to have achieved his object if he has turned over every stone, and solemnly described every strange creature that he may have found thereunder.

But a good review article, besides performing this archival function, should go much further. As I have emphasized, the primary literature is fragmentary, and only intelligible within the context of active research. It is a ridiculous, but commonly held, belief that the publication of the results of particular investigations is sufficient to create a body of knowledge. On the contrary; the information to be gleaned from a primary scientific paper is often about as meaningful as an entry in a telephone directory, or map reference in a military despatch; it only acquires significance by use, or by its place in a larger pattern, which must at some stage be made explicit. The job of the review writer is to sift and sort the primary observations and to delineate this larger pattern. He is neglecting his scientific duty if he does not offer criticism where it is justified, or if he fails to present a rational assessment of controversial matters within his province. It is only by such public reappraisals that those who are not already expert in the subject can have any idea of the credibility of the innumerable results 'reported in the literature'.

I cannot emphasize too strongly the importance of this activity of intellectual synthesis – a process of purification by recrystallization that must go through many stages, spread over many years, from the lecture by the chairman of a topical conference to the review article, monograph, textbook, encyclopaedia, and the *haute vulgarisation* of scientific journalism. Any notion that we may have about the nature of science includes the belief that something like an overall pattern is to be discovered and described. What we need is scientific knowledge – not more and more miscellaneous and unrelated information. The starting point for a search should not have to be an abstract journal or a computerized retrieval system – it should be an encyclopaedic treatise or textbook where the information has been transformed into an intelligible pattern of thought – if you like, a well-founded theory – from which can be deduced the characteristics of the particular datum, specimen, or phenomenon that we are studying.

Why is it so difficult to get this essential work done? It can be argued that the financial returns for such work are meagre, and indeed that scientific books will soon price themselves out of the market. This is nonsense. Scientific books are as necessary to research as scientific instruments. Nor do the royalties paid to the authors have much significance. The amounts are not large, but nor are they trivial when considered as a supplement to a regular salary. I do not believe that reviews and treatises can really be bought for hard cash from those who should be writing them anyway, and I should be reluctant to make the financial returns so high as to prostitute what must be, in the end, a labour of love.

Our present system of rewards and incentives in science does not encourage individuals to devote themselves for years on end to these critical synthesizing activities. But I am not going to propound some gimmicky scheme for prestigious prizes, review professorships, monographic fellowships of academies, and such like, to raise the status of the writing of treatises.

There is a coy reluctance to address oneself openly to other research workers in one's own, or neighbouring, fields, for they are supposed to know it all already. I appreciate that such writing is often ephemeral, and may quickly be overtaken by new discoveries – but that applies to all original research. There is a strange avoidance of explicit statement of basic principles. It is somehow deemed improper to set out the evidence, boldly, for and against some speculative theory. We are supposed, each one of us, to be carrying on our own internal debates on all such matters – and it is assumed that we shall all, by some magical dispensation of providence, thereby arrive at the same goals of comprehension. There is no appreciation of the great range of expertise and learning even among fully fledged professionals, so that a whole system of fundamental principles may be held as self-evident by one group, and yet be quite unfamiliar to others. We greatly underestimate the degree by which a set of new ideas may develop, in the minds of other people, with the possibility of revolutionizing our own point of view. At the research frontier, we are daunted by the very confusion and uncertainty of the knowledge we are to expound, instead of being challenged to make it clearer and more precise. Not only must we popularize our science to the layman; it is only by attempting to explain it to each other that we find out what we really know.

The trouble is, quite simply, a matter of philosophy. We are so obsessed with the notions of discovery and individual originality that we fail to realize that scientific research is essentially a corporate activity, in which the community achieves far more than the sum of the efforts of its members. It is not enough to observe, experiment, theorize, calculate, and communicate; we must also argue, criticize, debate, expound, summarize and otherwise transform the information that we have obtained individually into reliable, well-established, public knowledge.

References

1. Ziman, J. M. (1968) *Public Knowledge*. Cambridge University Press.
2. Garvey, W. D. and Griffith, B. C. (1964) Scientific Information Exchange in Psychology. *Science* 146, 1655–9.
3. Moravcsik, M. J. and Pasternack, S. (1966) A Debate on Preprint Exchange. *Physics Today* 19, 62–70.

37

Teaching Scientists to find Information*

How do scientists really find out what is already published?

Suppose one had to tell a graduate student how to find information on a subject with which he was not fully familiar; what would one say? My answer would be somewhat as follows.

This is a task for which there is no perfect mechanism analogous to looking at a clock to find out the time. Getting information is so central to scientific activities that one cannot prescribe a simple formula which will always be successful. Many sources are available; the choice of strategy will depend on the particular circumstances.

But the first step – and easily the most efficient if it is successful – is to find the *man* who really knows. Very often, he is someone on the premises – another member of your research group, a senior colleague of specially scholarly intellect, or some resident sage. He will tell you what really matters and you can have a helpful dialogue with him which ensures that he (and you) really understand the question whose answer you are seeking. There is no substitute for the personal *authority* of the expert. The information buck stops with him.

Unfortunately, the librarian is not much use as an authority on scientific information; he is, by definition, an authority on *books,* so his role comes later in this account.

Many people are so confident in authority that if they can't find one on the premises to talk to, they write to him. Unfortunately, the choice may not be obvious. For example, the man who wrote a book on the subject ten years ago is not necessarily still interested or informed about the latest developments. Again, it takes a bit of initiative to write such a letter, and an answer is not guaranteed. A passive technique is not uncommon: to wait for years until an authority happens to turn up at your laboratories and then spring the question on him. Actually, if there is no genuine

* Paper presented at meeting of the Abstracts Board of the International Council of Scientific Unions, Ustaoset, Norway, June 1972.

authority on the subject in your own country – this can easily happen in a developing country with a small scientific community – you are free to make up your own information; this has obvious advantages, since you can proceed with research to the benefit of your local academic career, though not perhaps in a way that helps your international reputation. This comment is not entirely a product of my own imagination!

But if a competent authority is not immediately available, you should look for a *book* on the subject. Here your friendly neighbourhood librarian may be able to help you, although a browse along the library shelf may turn up something. Of course you must *never* consult an encyclopaedia; the article may be written by the greatest living expert on the subject, and may seem perfectly clear, but of course he has had to simplify his account for the benefit of the lay public and therefore it cannot be trusted for real scientific information. Much better to turn to a standard work, first published 30 years ago, but available in a third edition of ten years ago. Here is the perfect written authority, a work whose ideas have stood the test of time and therefore must be true. Be careful, however, not to look into a second book, which may not agree entirely with the first. That would be confusing, and might make you doubt the information you have gleaned from the first book.

If there is no book, then you will have to be satisfied with a non-book, a volume of 'Conference Proceedings'. Thumb through this for the topic you want – ah yes, a very neat little paper (only ten minutes were allowed for its presentation at the meeting). In fact, it is so short and clear that it provides exactly the information you need. You can ignore the 'discussion' reported at the end of the paper; the main purpose of this is to satisfy the egos of those who intervene. It will not, of course, be possible to check the standing of the author of this contribution to the conference, although it may not be without significance that he is from the University of Timbuctoo. For reassurance, one can check that the results reported in the 1965 Conference do not differ substantially from those of the same speaker in 1963 and in 1961; they have stood the test of time.

The great advantage of a volume of Conference Proceedings is that it presents an outline of the state of the art in a particular field at that moment. One does not have to collect all the papers by all the participants out of miscellaneous journals. It is a sort of lazy man's current awareness aid, although normally retarded by five years.

If there is no Conference volume, then you must look for a relevant review article. This may not be so easy, unless there is some standard review series in the field which can be thumbed through to disgorge a suitable paper. Your librarian ought to be able to help find other reviews in more diffuse collections.

But by now you may have begun to worry about being reasonably up-to-date. You discover, alas, that the latest review was published in 1967 and cites primary papers up to 1965. The fact is that the editor of the review series recently asked the most reliable authority to write a new review or to suggest a suitable author, and got the answer that 'the subject was presently in such a state of flux that the time was not ripe for a review'. Alternatively, you may find 15 rather scrappy reviews of the same narrow topic in a variety of places, it being the fashionable band-wagon passing by. Fortunately, five of these are almost identical, being key-note addresses to various conferences by the same distinguished pundit, plugging the work of his own group.

The difficulty about gathering information from a review article is that it is customarily written in cryptic, veiled and jargonized language. It assumes that you know all about the subject already, and only need to have your memory jogged about some of the more obscure or absurd contributions. But what does it mean when the author says 'this point has also been discussed by McAlister, McWhirter, McGregor and McTavish (1957)'? Does one have to look up the *Proceedings of the Aberdeen Natural Philosophy Society* for enlightenment?

At this point you may well give up, and assume that nothing is known about the point you want answered. But wait, what about *Physics Abstracts?* The librarian will lead you to a back room, and you take down from the dusty shelves the latest volume. No, not quite, the six volumes from 1964 (more recent issues are being bound). You turn eagerly to the subject index. Yes, indeed there are no less than three pages, double column, under the single heading 'Ferro-electrics'. Which ones are relevant? You try to decipher the few words under each entry. Number 53972 looks hopeful, so you turn over the bulky volume. It turns out to be an experimental study on a few familiar compounds at the Bokhara Academy of Sciences; not likely to be very instructive. Anyway it will take a day or two to get it on Interlibrary Loan. Another item in the index is equally fruitless.

But in turning up this latter item, your eye is caught by another entry that looks relevant. Checking the original in the primary journal gives you a few other basic references which are cited as if significant. So you change your search strategy and leaf quickly through the appropriate sections in the monthly issues of the abstract journal, the current issues which have not yet gone to the bindery. In an hour or so, you have culled a few dozen primary papers in the core journals that are obviously relevant, and also a couple of reviews that were hidden away in unsuspected corners of the literature. It is not necessary now to search back into earlier issues of the abstracts, at least until you have assimilated the basic papers that really establish the structure and concepts of the information being sought.

By this stage you will have begun to realize that quite a lot of specialized services are becoming available to the earnest seeker after truth. Here again the librarian becomes your authority. It is his responsibility to know about abstract services, data banks, information centres and other devices which might be called to your aid. Lean heavily on him; there are occasions when the last ounce of information is worth its weight in gold.

Let me summarize with some general principles:

(*a*) Real *knowledge* is only to be found inside *people.*

(*b*) An ounce of critical *appraisal* is worth a ton of relevant rubbish.

(*c*) *Librarians* may not know all the information inside the books or other documents they store, but they can often be extremely helpful in suggesting where that information is to be found.

(*d*) Cryptic index items in a long list conceal more than they reveal.

(*e*) No machine can replace the searching *eye* linked to the discriminating *brain.*

One has to learn to glean information, even in the old-fashioned, browsing, inefficient way, just as one has to learn how to identify a geological specimen or to write a computer program. This is an essential stage in learning the art of research, which should be acquired during graduate school. It is the duty of the research supervisor, not the 'information specialist', to give this instruction, preferably by example and confirmed experience.

38

Whistle-blowing*

The critical eye of the referee is the lynch pin of the scientific enterprise.

What's this big envelope from the Institute? Not *another* paper to referee? I'm too busy. They know I'm too busy. Why do I get all this junk? Can't they send it to somebody else? Who's it this time? Oh, *no!* Not *Bloggins* again! The man's a fool. Everybody knows he's a fool. Sheer nonsense of course. I don't know why they go on publishing his stuff. The referees are much too kind; it ought to have been rejected long ago. Well, they made a mistake sending it to me. I'm not frightened of him. I'll go through it line by line, and really take it to pieces. And I won't mince my words. He's not a research student now. He's got to learn proper standards some time. A bit late now, with that Personal Chair, but that's their mistake. I must just tell the truth. That's why referees are anonymous, isn't it – so they won't be frightened to tell the truth.

Now, how to start; 'The theme of this paper, although superficially plausible, is based on unproven assumptions that must now be regarded as false. The author seems quite unaware, for example, of the work of Coggins (1972) . . .' Yes, that sounds about right. Got to blow the whole thing up. No good niggling about details. Anyway, he usually gets his algebra right – just doesn't know how to apply it. But that reference to my paper is a bit obvious. Mustn't give the show away. He mustn't know it's me. He'll be at the SRC committee next month, and we'll have to stop the chemists. Anyway, he's perfectly decent as a chap: not like Doggins, who's always thinking of number one. Pity old Foggins didn't treat him harder when he was a student – just not critical enough, that's the trouble.

Well, let's look at his references. Yes, just what I expected: Foggins & Bloggins (1963); Bloggins & Foggins (1965); Bloggins (1966); Bloggins, Hoggins & Joggins (1970); Bloggins & Joggins (1971) – the same old stuff. But what's this – Coggins (1972)? So he *has* read it after all.

* Leader in *Physics Bulletin* 25, 9 (1974).

But he won't have understood it. I bet he says something like 'Coggins (1972) has failed to appreciate . . .' and brushes it aside. Here it is, on page three. *Good Lord!* 'It now seems clear from the work of Coggins (1972) that the only satisfactory solution to this problem must be based upon the following assumptions . . .' Perhaps he's solved it after all! Who would have thought it? Good old Bloggins: I always said he was clever. Ah well, that makes it easy. Not often one gets a paper to referee that one thoroughly approves of. Scarcely need to check it – he'll have got the algebra right. Let me see; how shall I start? 'This admirable and original paper makes a considerable contribution to our understanding of . . .'

39

The Light of Knowledge: New Lamps for Old*

What sort of information system will be needed by the scientist of the future?

It is impertinent, as a mere consumer, to be addressing a gathering of professional experts in the communications industry. I feel rather like a housewife being invited to speak to a group of electrical engineers, or to a conference of managers of municipal water supplies, about the product that she expects to flow when she presses the switch or turns on the tap. We just want good clean information, at a standard voltage, well filtered, safely insulated, and ready for use; that is all there should be to it.

But of course this metaphor breaks down – we scientists are as much the producers of knowledge, as the users. At the risk of being indelicate, I might equally well regard you as sewage disposal experts: there are moments when the fashionable phrase 'pollution of the environment' seems an appropriate epithet for the vast quantities of waste words that flow out of the scientific community and require to be filtered and purified before the information they contain can be put to use!

At last year's meeting of the British Association at Exeter,[1] I tried to give an account of the communication system of science, and to explain its rationale as a social organization. So much has been preached and prophesied against our conventional system that it seemed useful to make explicit the function that it is designed to serve and the reasons for some of its peculiar features, before we decide to take it all to pieces and build a far, far better world. Even if one does not believe that All is for the Best, that the *ancien régime* must at all costs be defended, that More means Worse, and so on, it is a good exercise to put oneself into the shoes of the conservative or reactionary, and to play the advocate in his cause.

My client this evening is a somewhat different character – younger, less stable of purpose, more at the mercy of his emotions, lacking in worldly

* Fourth ASLIB Annual Lecture, delivered in London, April 22, 1970; published in *Aslib Proceedings* 22, 186–200 (1970).

experience, but full of energy and zeal. I propose to think aloud on behalf of the scientist of the future. What sort of information system will he devise, instinctively, to suit his needs? Since we know nothing for sure about this hypothetical person, this lecture is pure speculation; but science fiction may be as effective as history in the spinning of fables and the manufacture of myths. After all, the transformation of society by science is quite as magical as any of the feats of the Djinni of Aladdin's Lamp.

First let me draw attention to three characteristics of the orthodox system of scientific communication, as we may find it in the learned journals, treatises and handbooks that line the bookstacks of your great libraries. The primary function of an ordinary scientific paper is to bring into the public domain an explicit account of some new scientific development or discovery. It is a rather contrived document, very conventional in form and style, whose purpose is to persuade the general scientific community that the author's new observations or arguments are of interest, significance and permanent validity. As I have argued at length in a recent book,[2] the whole activity of science is dependent upon the publication of such documents, and their subsequent criticism, reevaluation, and eventual acceptance or rejection as parts of the consensus of 'Public Knowledge'.

But scientific publication also plays an essential role in the social system of science as the registration of priority of discovery. Some sociologists of science, influenced by current fashions in anthropological theory, go so far as to describe published papers as the 'gifts' brought by the research worker to his community, for which he is rewarded by recognition and esteem.[3] This interpretation exalts to a somewhat mystical plane our hard-headed practice: in the agony of choosing between worthy applicants for academic posts, we look carefully at the publications of the candidates, in the hope of distinguishing therein some more objective measures of intellect and imagination than are vouchsafed to us by the character references of their previous employers. In other words, since research is the production of new, useful, 'consensible' knowledge, the original papers of a research worker are the tangible evidence of his ability to do research. Surely, if one were appointing a potter to make beautiful pots, then one would inspect the pots he had already made, in the hope of finding them to one's liking.

Finally, I would draw attention to a uniform characteristic of the whole conventional system: the *user* of the information pays all the costs of publication. Books and journals are published by learned societies and commercial organizations, and sold for profit – or at least to cover all the printing costs – to individual research workers or to the organizations for which they work. The library is a cooperative of *readers,* not of authors. This is a point of some principle for it is the ultimate source of the power of the scientific editor or referee; he may reject a contribution on the grounds that it will lower the standard of his journal, or monograph series, and hence endanger its financial viability. As I have argued at length,[1,2] the function of the referee is essential if the scientific literature is to be kept free of crankiness, irrelevance and gross incompetence. A scientific library, also, is selective in its choice because it must keep within its budget, so that one can feel confident that the books one turns to for reference are the most reliable available, within the judgement of the librarian and his technical advisers.

This conventional system, however well it has worked in the past, is now in a state of crisis. This crisis is not, I would submit, an internal development of the system, but is due mainly to the enormous expansion of scientific activity in the past few decades, putting too heavy a burden on organizations and procedures designed for a more leisurely age. It is not necessary for me to quote figures for this expansion of science, nor to flash onto the screen one of those startling graphs, drawn on a logarithmic scale, showing that if the present rate of growth were to continue unchecked, then by the year 2500 every man, woman, and child on Earth would be doing nuclear orientation experiments, or contributing papers on algebraic topology, or all simply tapping away at the teleprinters in the effort to keep up with the computerized indexing of the abstracts of the papers that would by then be pouring off the presses like Niagara Falls. If you do not have this piece of general information firmly in your minds by now, then you have no right to be present at this lecture.

Let us look at this crisis from three different aspects. Much of the discussion has concentrated on the difficulties of indexing, abstracting and assimilating the vast quantity of new material, proliferating from many breeding grounds and overrunning the poor scientist who is just trying to keep up with the literature. This seems to me to be largely a technical problem that can be solved quite effectively if we are prepared to put sufficient resources into it. It is not obvious, to my mind, that a 'technological fix' with arrays of half-witted computers is the only answer: there is still plenty of scope for development in the old-fashioned procedures of compiling, indexing and printing up-to-date collections of abstracts and reviews. In any case, this is the sort of subject where the consumer has no rights; I leave all that to the expert designers of generators and turbines, dams and pumping stations.

From my point of view, a more intractable consequence of the expansion of science is the debasement of the criteria for quality of publication. Not only is there too much scientific work being published; there is *much* too much of it. I do not wish to argue, at this stage, that the scientific potential of the new generation of research workers is significantly lower than in the past, but simply that the need to get recognition by publication forces each of us to shout a little longer and louder so as to be noticed at all in the gathering, swelling crowd of voices. If we cannot get into the quality journals, we send our pitiful contributions to a second-rate publication, produced perhaps by a commercial publisher who has learned that vanity will get him a distinguished-looking list of editors and the professional zeal of librarians will ensure that he gets his money back. Or we may simply have our work printed, or duplicated, as a 'Technical Report' or (hopefully) as a 'Preprint', and send it round free to everybody who we think ought to be interested. The effect is that the referee system is being eroded at the edges: however hard we try as editors to keep up the standards in the central, well-established, respectable, archival literature, we find ourselves on an island in a vast sea of semi-literate, semi-scientific, half-baked and trivial material, which threatens to swamp the whole system.

Another serious consequence of the expansion of science is a steep rise in the cost of our information services. If we conceive of an information system as a network of pathways from individual to individual then N individuals generate $\frac{1}{2}N(N-1)$ separate channels of communication. Thus, as N increases, the number of channels required by each individual also increases. If the bulk of literature in his subject is growing, then it costs him more to purchase all the journals and books he needs, whether from his private purse or through an institutional library. For the individual scientist these costs can be kept down by narrowing his attention, and only purchasing material within his own specific field; for example, instead of taking the whole of the *Physical Review* or *Proceedings of the Physical Society,* he may now buy only the sections devoted to Elementary Particles, or Nuclear Physics or Magnetism. But for institutions such as universities, this economy is not available, and scientific library costs are really beginning to hurt. In Western Europe and other rich countries, this extra burden can be carried; but for many developing countries, and especially those states with pathological objections to the expenditure of foreign exchange, scientific progress is now hobbled almost to a standstill by the inability to buy the full range of publications that now contain the modern stock of scientific knowledge.

Indeed, the attempt to solve the information crisis by computerized retrieval aggravates the position of a university or research institute in India, Africa, or Latin America, for whom these gadgets are much too expensive. I do not know whether anyone has actually calculated the cost of such devices per research worker using them, but it would be tragic if they came to seem indispensable. The disadvantages of the humble scholar in Chittagong, Cape Coast, or Belo Horizonte are desperate enough, by way of inadequate apparatus, and intellectual isolation, without adding this further cause for feelings of hopeless inferiority and poverty.

It is my impression, however, that discussion of the information crisis has seldom been accompanied by a recognition of the rapid structural changes that are taking place within the scientific community which the communication network is supposed to serve. The tacit assumption of the librarian, journal editor, information officer or scientific publisher is that he will always be dealing with the products and needs of a vast horde of individuals – that he is a broker in a peasant market, where pigs and sacks of corn, cart wheels and chairs, are traded from hand to hand, for cash or barter.

It is obvious, however, that scientific research is rapidly evolving from a cottage industry to a system of factory production. This, surely, is an inevitable consequence (if not, also, a cause) of the enormous expansion of science in recent years. If an agrarian population were to grow tenfold in a working lifetime, it could not be accommodated by scattering ten times as many little huts and allotments over the same fields; towns and cities, tenement buildings and office blocks, supermarkets, railway stations, government secretariats, and other appurtenances of high civilization would have to appear. Is this not the appropriate simile for those vast hives of scientific buzzing and honeymaking that we now observe throughout the world, those national laboratories, research institutions, institutes for atomic energy, or for cancer research, or for apple growing, or for seaweed, or for noise, or for space, or for higher thoughts, or for lower temperatures, or for advanced studies, or for retarded minds, or whatever it is. What are they but *factories* for the production of knowledge? Compared with Brookhaven, or CERN, the Cavendish Laboratory of J. J. Thomson and Ernest Rutherford was as a blacksmith's shop to a steel mill. Faraday and Edison, Pasteur and Darwin – they were individual artists or artisans, with perhaps a few apprentices as extra hands; their successors are executives, captains of industry, directors of research, managers of departments, chairmen of research councils, and other high-powered bureaucrats.

There is not the time here to analyse this development. To some extent it flows from the demands of Big Science – the need for the combination of resources and effort around some especially powerful and expensive instrument such as a particle accelerator, telescope, rocket launchpad, nuclear reactor or computer. It is a natural tendency in applied research and development, where vast sums are made available for a concerted attack on some practical problem – a nuclear bomb, radar, a new telephone system, anti-missile missiles – without regard to the extravagance of concentrating too many minds, in too short a time, within such a limited intellectual sphere. There are economies of scale, in the provision of common services such as libraries and computers. There are natural psychological tendencies; many people enjoy the comradeship of others working on the same problems, and are stimulated by the *schwärmerei* of the big buzzing group. There are factors of efficiency and tidiness in the minds of the administrators of government grants, who see virtues in focusing their charitable offerings on the centres of excellence, on the proven scholars with their large and productive groups, rather than spreading a thin layer of bare sustenance over the whole academic community. It can be argued, with a good deal of plausibility, that the relatively modest talents of those nine new scientists in our generation are best employed under the direction of the tenth, who has the inborn ability and imagination for creative research – that only by a division of labour and a team system can we really make scientific progress in reasonable proportion to the numbers of persons involved. Whatever the reasons, good or bad, the fact of industrialization and bureaucratization of science remains.

The consequences of this historical development for the life of the mind have not been seriously discussed; perhaps we cannot discern the effects ourselves, as the wave breaks over our heads. But I would draw attention to two obvious changes in the personal status of the scientist: he becomes more professional, and more subordinate to a team activity.

There have been 'professional' research workers in Western Europe for several centuries – the academicians of St Petersburg and Paris, supported by the State as full-time scholars. Nevertheless, until quite recently, the bulk of scientific research was done by university teachers, not as an explicit part of their employment but as a semi-independent, personally directed activity, motivated by the desire for further preferment as a reward for scholarly prowess. This pressure could be agonizing[4] – witness the bitter struggles of Sigmund Freud to achieve recognition for his discoveries – but it was indirect and unspecific. The official doctrine was that a university lecturer, being called to the vocation of high scholarship, would naturally wish to add to the corpus of pure learning, and was given the freedom to make the best of his time in that way. Within the bureaucratic institutions of our day, this amateur status has been renounced. Whatever freedom he may have in the day-to-day, month-to-month, even year-to-year direction of his research, the modern scientist is none the less paid to be a research worker, and is expected to produce some results of a publishable or useful kind, in return for his stipend. He has not really the option to retire from active scholarly investigations, and to devote himself entirely to teaching, or other duties within the academic community. Like many a Gentleman, he has been handsomely bribed to become a Player.

In the best research institutions, an effort is made to preserve the freedom of the individual scientist in his choice of problem and in the tactics of his research. Nevertheless, the tendency towards team work grows ever stronger. In Big Science,[5] the experiment, and the discovery that arises from it, is not a product of the ingenuity of a single mind, but of a large group, whose members specialize on different aspects – setting up the beam of particles, arranging the counters and spark chambers, designing the electronics, writing the logical programs, checking the overall behaviour of the system, watching the dials as the experiment proceeds, analysing the data, and finally interpreting the results. The evidence for this multiple effort is the multiple authorship of the paper in which the results are reported – a dozen, a score, even half a hundred names appended to the publication. The individual skills required may be of the highest order – but the 'productivity' or 'creativity' of each participant is merged into the collective mind of the team. An investigation that costs a million dollars is clearly a large-scale social activity, not at all the same thing as a personal work of art for which credit may be assigned to the individual creator. The war against ignorance is fought by well-drilled companies and battalions, not by heroes in single combat. Success may depend as much upon good staff work, sound logistics, expert training, and skilful command as upon individual courage and skill. The office desk – the *bureau* – takes precedence over the laboratory bench; intelligent and precise execution of orders may be more valuable in subordinates than independence of mind and untrammelled imagination. The very idea of tussling with a problem of natural philosophy may be forgotten in the struggle to achieve some immediate technical aim within the vast framework of the apparatus.

Now let us ask ourselves what effect this change in the social and psychological structure of science may have on the information system. It is obvious, in the first place, that there is a grave threat to the convention of awarding promotion, or other forms of recognition, on the strength of published work. The mere fact that a candidate for a lectureship in elementary particle physics has his name amongst the dozens of 'authors' of some significant discovery says little about his scientific skill. In the long run, the leader of such a team gets the credit for its contributions to knowledge, but he must be already the selected and tested boss of a big group. Evidence of ability at a more junior level can only be assessed within the framework of the project itself, just as it would be in an army, a civil service or other bureaucracy. This is not the place to discuss the psychological consequences of this development, which puts direct power into the hands of the seniors, and opens the way to careerism, personal autocracy and other evils, as well as giving the advantage to the 'other-directed' personality, at the expense of those Protestant virtues of being 'inner-directed' which have contributed so much, in the past, to the scientific attitude. All I need to say here is that one of the primary functions of the conventional communication system of science is losing weight. The necessity of maintaining an open market for the creations of the individual scholar, as objective evidence of achievement and promise, is no longer evident.

On the other hand, the funding of research on a vast scale, the assumption that every scientist is a professional, employed and supported materially by a large-scale institution, makes it easier to find the money for expensive new techniques of communication. We begin to take it for granted, for example, that we may charge to laboratory overheads our long-distance telephone calls, from Berkeley to Brookhaven, from London to Geneva, from Trieste to Washington, which we would never dream of paying from our private pockets. Air mail for reprints, air travel for conference participants, secretarial assistance, Xerox machines, electric typewriters, photo-offset printing, all are enlisted in the battle for more and more rapid, total, instant, informal communication. It is easy to prove that such aids, although not of negligible absolute cost, pay for themselves many times over within the context of million dollar research budgets, palatial buildings, gargantuan machinery and armies of assistants.

The question that has not been resolved is how this money is best spent in the support of the *formal* communication system. We find, in practice, two divergent trends. On the one hand, the provision of more and more lavish library facilities is taken for granted, along with the new gadgets for information retrieval and search. As I have already remarked, the affluence of science in the more industrialized nations can easily keep pace with the flood of new publications. The assumption is made that library funds will always be increased to permit the purchase of all the new periodicals and books that are relevant to the research on hand, with lavish allowances for treasures that may still be unearthed in the older literature. The recognition that these costs are really trivial by comparison with overall expenditure on buildings, equipment, technical staff and professional salaries was the secret of at least one great fortune in the technical publishing world – and it remains largely true today. Some of us have jibbed at the enormous expense of such marginal aids as the *Science Citation Index*, and we are worried at the high prices now being charged for advanced treatises and other books, but we can still maintain an open-ended commitment to obtain all the genuine literature of our discipline without making deep inroads into our overall research budgets. Thus, the financial power of bureaucratic science is put at the disposal of the users of information; the larger the scale of the laboratory, the easier it is to provide such facilities on a collective basis.

On the other hand, this financial power is being applied to the information system at the *producer* end. I have already referred to the growth of the 'Report' literature, especially within the very large government agencies and industrial laboratories. This material may not be produced in the full glory of letterpress, but the retyping and duplicating of a few hundred copies of a 20-page report or preprint, not to mention air-mail postage and incidental overheads, is not a negligible expense. The fact that it is distributed free – that it is almost forced upon one, like an advertisement for laboratory equipment or life insurance – is in striking contrast to the high subscription one has to pay to read the same article, only a few months later, in a regular journal. The pressure to get into print, for professional prestige and preferment, is thus diverting a substantial amount of research funds into the information system, but in an extravagant manner that is largely dysfunctional to the scientific community, through the degradation of standards and the clogging of the channels by multiple publication of the same basic research.

A parallel development is the rapid growth of the page-charge system. It is an ancient custom for a journal to supply offprints of an article to its author, at a run-on cost, and this is not begrudged as a service to friends and colleagues who might find it convenient to have a good printed version of the article ready to hand. But this has now been inflated into a really steep charge – $80 per page in the *Physical Review* – levied on the institution of the author when the paper is accepted for publication. It is argued that publication is part of the overall research activity, and therefore its cost should be met out of the research grant, or other fund, that pays the salaries and equipment of the researchers.

This is perfectly logical; but it does conflict with the general principle that publication of scientific information is paid for – hence, eventually controlled in quality – by the reader. The actual pressure behind page charges is the desire of each editor to keep the selling price of his journal within the means of the individual subscriber. This is particularly so for the great scientific societies, whose members have become accustomed to receiving their journals at quite a modest price, as a special concession and a service. I will myself testify that I belong to the American Physical Society, not out of international brotherhood and solidarity, but because I can get a great deal of valuable scientific material – several thousand pages a year – for a subscription of only $50 or so. It is a good bargain, and would be particularly significant if I were working in a developing country where library costs were a real burden.

Nevertheless, page charges become a tax on research, and can only be borne by heavily bureaucratized science, where the employee can assume that somebody else is picking up all the checks. The small research group, still trying to produce in the cottage industry mode, gets squeezed out. Is it really the intention of the American Institute of Physics to prohibit in its journals any publications from the outside, dollar-less world? I certainly do not propose to spend the salary of a good research assistant on having the papers from my own research group published in the *Physical Review,* when I get them printed for free in the corresponding British journal. The AIP say, of course, that the charge is not obligatory; but they are now delaying publication of papers from 'irresponsible' institutions that do not 'honour' their page charges, and have begun to ask higher subscriptions from foreign members on the grounds that we are not paying our fair share of the printing costs. In other words, we are approaching a situation in which the whole scientific community of the United States is regarded as a single body, supported financially and thereby controlled by the major funding agencies of the Government. This support is not a disinterested gift to learning and to humanity, but is evidently calculated as the cost of buying certain specific services for the Nation.

As I have said, I do not propose, in this lecture, to discover conspiracies, nor to denounce the more sinister aspects of the world we live in; in any case, the brutal bureaucratization of science in the Soviet Union and its colonies is far more tragic and foolish than any situation they can manage to bungle their way into, through sheer expediency, in the

United States. The significance for the information system is that the institutions that employ the scientists and treat them as a category of skilled manpower, hiring and firing them by hundreds, throwing them into battle as shock troops or keeping them as reinforcements in reserve, counting them as realizable capital assets, as 'Human Reserves' in the balance sheet – these institutions are now laying their hands on the information system, and beginning to take charge of the various mechanisms for the publication of knowledge.

Just at present, we are in a state of transition. The individual scientist – or at least the leader of the research group – is allowed to say what he likes, how he likes, and takes personal responsibility for his research results just as if he were an old-fashioned professor. He may, indeed, have at his disposal the funds to advertise his work through the distribution of reports and preprints, without any critical check. On the other hand, he must still submit his results to the scrutiny of independent anonymous referees if he wishes to see them published in a respectable journal. The financial power of the laboratory that employs him is not used to interfere significantly in the form and substance of his published work.

How long can this situation continue? Shall we see, before long, a tightening of the strings within the bureaucratic organization, so that the decision to support publication will rest with higher authorities than the man who really did the research? In a new historical phase where money has suddenly become tighter for research, it seems inevitable that more careful control of page-charge, report, and reprint costs will be exercised, so that only work judged worthy by the senior officers within the institution will be cleared for publication. To anyone familiar with scientific controversy, and the self-righteous malice of the second-rate senior scholar found in the wrong, this is an unhappy prospect. The history of Lysenkoism[6] shows the destructive effect of ignorance and incompetence when it gets its hands on the levers of power of a complete bureaucracy, and controls publications as well as employment. We could see a covert conspiracy of the publishing monopolies with the scientific institutions, in which awkwardly critical or uncomfortably heterodox research could get no public hearing.

I am not quite so pessimistic as this. I suspect that the institutions will take firmer control of the publications of their employees, and insist on higher standards of style and of achievement before they will pay the bills. In other words, the published work emanating from a particular laboratory may be seen more and more as a product of that laboratory,[7] rather than as the creation of an individual scientist who only works there. As it becomes more difficult to make a name for oneself by one's personal list of publications, the pressure to put on the market vast quantities of half-baked material may recede, and a longer rhythm of work, with less frequent, less fragmentary papers, may succeed. Of course, there will still be tremendous competition for priority of discovery between groups in different laboratories, but this may well be catered for through the informal system of letters, preprints, etc. The actual *number* of papers ceases to be of much significance as a standard of productivity by a substantial institution.

We could arrive at a situation where in practice the role of the referee, as a professional critic, would be performed mainly within the producing laboratories, or even within the research team itself. It must be remembered that the function of the editor of a journal, or of his anonymous referees, is not to guarantee the complete scientific validity of every paper that eventually gets published. One only finds this notion in the minds of certain notoriously hypercritical scholars, who eventually get dropped from the panel of experts! The job of the referee is to confirm that the work is scientifically interesting, more or less original, reasonably well expressed, and not vitiated by obvious errors that the author can be persuaded to acknowledge. It provides a barrier against irrelevance, crankiness and gross incompetence – not against mild stupidity or subtle mistakes. These conditions can be met quite easily by the informed critical examination of work in progress within a capable team or well-established laboratory. No doubt there are occasions when the leader of a research group imposes his personality so strongly on his juniors that he can persuade them to put their names to work that they do not quite believe in – but we also have examples where the name of some famous scholar is sufficient to blind the anonymous referees to the weakness of his argument. In other words, I am prepared to believe that the standard of work published at the behest, and in the name, of a well-established laboratory or other scientific organization might be somewhat higher than the average produced by individuals in the current literature.

Shall we then see a new system, in which the great scientific corporations announce their discoveries in 'house journals', edited internally and distributed for free? In general, such a system would not conflict with the fundamental characterization of science as a community devoted to the creation of 'consensible' knowledge. Given reasonable freedom of speech and action within the corporate institutions, critical comment and the eventual arrival at a rational consensus would still be achieved. The 'scientific attitude' has become so ingrained in the thinking of a generation of professional research workers that even such power structures as Lysenkoism can be defeated by an appeal to simple fact and straight logic.[6]

But it seems absolutely essential for the preservation of this attitude that the bureaucratic corporations should not monopolize the sources of information and means of communication. One single international institute of high energy physics, dominated by its trillion volt machine and controlling all the media of publication in this field, would be a monstrosity. Competition for priority between individuals must be transformed into intellectual controversy between coequal groups and teams, with public debate and open criticism the major weapons.

It is particularly important that the old-style journals, published at the expense of their readers and preserving high standards through the use of anonymous referees, should not fall for the expedient device of the page charge. As I have said, there is a danger that the learned societies may fall in with the bureaucratic trend, and fail to protect the rights of those of their members who still have freedom to do research on their own account. Can't you just see it? A paper arrives from a private address. 'Don't know this name . . . Doesn't seem to be in the directory of Ph.D.s . . . Where does he work? Seems to be doing some desk job in the Patent Office . . . They don't have a program in astronomy, surely. Has he got permission to publish from his research manager? I bet they won't pay their thousand quid for this stuff. I'll call up the Director: what was the initial? Oh yes: A for Albert – OK, I've got it. Albert Einstein.'

To maintain the older system of scientific communication within the framework of the new bureaucracy and the new technology will not be cheap; to many, it may seem no more than the preservation of a charming traditional custom. The new procedures which I have sketched out will have many advantages, permitting extensive rationalization of the information network, and much easier access to libraries, computerized indexes and other means of information retrieval. Whatever is published by the great laboratories and other corporate bodies may be better written, better printed, more uniformly of scientific interest, and better indexed than anything we have known in the past. For *normal* science – the solving of puzzles within an accepted framework of ideas, by well-organized teams, using expensive apparatus, under the direction of the experienced professionals, this may be the best way to do it. But the mechanisms for revolutionary transformations of thought such as those generated by a Darwin, a Mendel, a Wegener, a Planck, have always been frail, and must be preserved at all costs. I am sorry for the Indonesians and Bolivians, unable to buy all the journals they need to keep themselves scientifically informed, but I do not think it would be to their benefit if the world's scientific literature were given away, like advertisements for patent medicines, yet all doors were then closed to the publication of their own modest researches.

I have left to the last the very important problem of reviews and books within the framework of big, corporate science. As I have already suggested, the supposed crisis in technical book publishing, with hard cover or even paperback editions pricing themselves out of the market,[8] is not the real issue; the difficulty is to get them written in the first place. There is already a serious lack of incentive towards review writing, instead of primary research,[1] fostered by a false philosophy of originality and 'creativity'. It is held to be a waste of one's talents (however modest!) to 'take time off from one's research' for the supposed drudgery of this 'derivative activity'. Even within the conventional system, the writing of books and review articles has become intellectually devalued, and no longer regarded as a high art of scholarship, contributing greatly to the furtherance of knowledge and winning laurels of fame for the author.

Of course, the financial rewards are usually quite modest – only those few authors who were cunning enough to fill a gap in the booklist of the half million college freshmen in Sociology I, or European Civilization 001, or even Algebra 1A, have made fortunes in this business. For the ordinary good scientist, with an excellent monograph or treatise waiting to issue forth, there is not a hope of his royalties covering even his normal income for the time spent in writing, let alone all the years of experience and reading that have really gone into it. If scientific books were composed by professional writers, then we should have a depressed class of hacks on our hands, as poor as would-be novelists and infinitely duller. But, of course, most reviews and books are written by people already enjoying a full academic or other institutional income, and the payments they receive for this work are additional 'perks', which help to get the house painted or to buy a sail boat. The marginal financial incentives, generated within the conventional system of cottage-industry science by the thirst of the readers for great draughts of fresh new knowledge, are not negligible, but they are not sufficient to buy the best in authorship.

The bureaucratization of science weights the balance even more heavily against the writing of reviews and books. It is psychologically much more difficult for the member of a team to withdraw himself from active experimentation than it has been, in the past, for the lone individual. Imagine yourself, once more, in a group devoted to high energy physics, or to space research. The experiment has been planned several years previously, as part of a complex continuing program, where many groups must take their turn at the receiving end of the accelerator or satellite. The technique of observation, the design of the apparatus, the testing, proving and running of the equipment, are all novel, and only marginally reliable. The whole process, spread over many months, is a race against time, against the schedule of use of the beam or rocket, and against rival groups. There are continual crises, and conferences, and unplanned modifications to develop and test. Skilled manpower is in short supply, and everybody is driven on, relentlessly, by the ambition and enthusiasm of the leader of the team. How can one possibly withdraw from such purposeful activity, with its clear and immediate goals, just to sit, and think, and write a few hundred words, and scratch it out again, and wander aimlessly round the library seeking inspiration, and go back to the desk, and try again to express with due clarity and simplicity the convoluted thought one has nearly mastered in another man's work. In the busy workaday world of making, mending and doing, this looks like *dolce far niente,* or even treason: 'We've no room for passengers on this craft' says the leader of the team – handing one a monkey-wrench, and indicating the vacant place around the apparatus where it is to be applied. If one is trained from manhood to work in such a team, one can scarcely abandon one's friends and colleagues in such circumstances.

Indeed, if one is paid full-time to do research, then, by golly, research one ought to do. 'Research', in the eyes of most managers and grant givers, is what leads to primary publications – real new discoveries – not this messing about writing books. Thus, even the individual lone scientist, employed by a university or other institution, may come to feel that he is not really earning his salary if he spends his working time in this apparently unproductive activity. As can readily be shown, by an appeal to the general principle that science is *public* knowledge, necessarily subject to a continuing process of critical assessment and reintegration, the activity of the review writer is quite as important to the scientific community as the making of 'original' discoveries – but the purpose of the bureaucratic research corporation does not extend that far. We face, indeed, one of the familiar paradoxes of society – the 'tragedy of the commons'[9] – in which the immediate self-interest of the individual is best served by his adding another goat to the grazing herd, rather than by fencing and controlling the use of the common land. In so far as the individual becomes the paid employee of the purely selfish and soul-less corporation, he loses his liberty to act in an enlightened way in recognition of the needs of the community. In other words, the shallow philosophy of 'getting on with one's work' – that is, piling up research findings – is actively reinforced by the attitudes of management, dominated by quantitative measures of 'productivity' and 'creativity', and not caring at all about the general health of the art.

Review and book writing is thus driven into the 'spare time' of the scholar – evenings, weekends and occasional sabbatical leaves. Even in the university it is not considered proper for any scientist except a professional 'theoretician' to spend his ordinary working hours in this way. There could develop a threat to the modest financial rewards to be won by this extra effort – high taxation on additional earnings, or the demand that 'outside' income by holders of academic posts be disclosed and curtailed in the name of equality, fair shares and maximum productivity. One can just see a certain type of Treasury character arguing most persuasively that since book writing must constitute a diversion of effort from the teaching and 'research' for which each Professor and Lecturer is already paid in full according to his exact merits, the royalties should be counted as a contribution to the funds of the university itself (the corresponding amount to be deducted from the UGC grant, on receipt of full returns countersigned by the Registrar and Vice-Chancellor) and not go into the pocket of the employee who happened to undertake this irrelevant but lucrative activity.

Well, we have not quite got to the point yet – though I see no impenetrable barriers against it within the basic axioms and theorems uttered on behalf of most of the corporate bodies in the spheres of government and industry. We must evidently begin a campaign now, before some sort of disaster of fragmentation falls upon the whole system of communication in science. A number of lines of action suggest themselves. The providers of research grants, for example, might consider much more seriously the explicit subsidization of such writing, by providing funds for sabbatical leave that would free a university teacher from his everyday duties for a certain time, not to do more experiments but specifically to write. There have been occasions when the writing of a book has been proposed as the goal of a research project, but from what one hears this is not usually received with favour by the expert advisory committees, who perhaps set much higher standards for what they regard as the essentially mechanical work of surveying what is already known than they do in estimating the likelihood of success of a proposed new investigation. They must be made aware of their own unconscious belief that writing is really much more difficult and dangerous than experimentation (for it exposes the mind and soul to criticism and ridicule) and therefore be lenient and generous towards those who are willing to give it a try.

Within Big Science, it may be necessary to assign the role of 'scribe' or 'theorist at large' to a particular member of the team, or to provide income and facilities for a number of such experts within the institutional framework. Perhaps this is already happening – for example, in the enormous space-science corporations and administrations. But notice the danger of specializing and professionalizing this particular role. The knowledge and learning needed to write a book comes only by direct experience at the research front, by participation in the uncertainty of experiment and argumentation. No amount of external reading around a subject can give one the feel of it; a particular field of science is like one of those vast, extended, African or Indian families, into which one must be born, or to which one must be wedded for many years, before one can understand or explain what it is really about. Thus, every research scientist ought, from time to time, to put on for himself the mask of the reviewer, to see his subject whole and express his own wisdom; he cannot leave this essential task to a hired writer, however expert at grammar and syntax. The invitation to assume this persona could as properly come from his research manager or other bureaucratic superior as any instructions concerning research to be undertaken or classes to be taught.

Finally, the new tendency for the cost of the information system to be met by the producer rather than the reader might be allowed to take effect here too. The books written by those employed for that purpose by corporate bodies could be produced cheaply, and heavily subsidized, in the name of scientific progress. If the incentive of royalties and other personal rewards is still to be active, then a substantial subsidy to non-profit publishers might make it much more worth while for authors to stop research or teaching for a while. The bookwriting world, fiction and non-fiction, has such a peculiar economic structure, and seems to defy so many of the theorems of the market place and factory, that one cannot predict just where and how the extra funds would have the best leverage – but it seems to me essential that corporate action be taken to restore to the scientific community as a whole these services that are now threatened.

In this lecture I seem, once more, to have taken a rather gloomy view of the future. But of course it is the duty of the social commentator to 'view with alarm', and predict catastrophes that might indeed occur if one did not take evasive action; pessimism is as proper to his tone as gravity is to an undertaker or frivolity to a debutante. After all, the worst does not always happen, and we can have pleasant surprises. For all the scale and complexity of its social organization, for all the crimes and follies committed in its name, there is a certain simplicity and purity about science, inherent in the image of the curious man studying the world about him, and finding it interesting, intelligible and good to look on. In the Tale of Aladdin, the Princess was tricked into exchanging the magic lamp for a new one that would not work. I still have confidence that we shall be able to call the Djinni to our service, nearly as well with our new lamps as we have with the old one these last three centuries.

References

1. Ziman, J. M. (1969) *Nature* **224**, 318.
2. Ziman, J. M. (1968) *Public Knowledge*. Cambridge University Press.
3. Hagstrom, W. O. (1965) *The Scientific Community*. New York: Basic Books.
4. *See* Gerth, H. H. and Mills, C. W. (eds.) (1946) *From Max Weber*, p. 129. Oxford University Press.
5. Swatez, G. M. (1970) *Minerva* **8**, 36.
6. Medvedev, Zh. A. (1969) *The Rise and Fall of T. D. Lysenko*. New York: Columbia University Press.
7. Witness, for example, a paper authored by 'Orsay Liquid Crystal Group', *Phys. Rev. Letters* **22**, 1361 (1969).
8. Benjamin, C. G. (1968) *Scientific Research*, 16 September, p. 32.
9. Crowe, B. L. (168) *Science* **166**, 1103.

PART EIGHT

EDUCATION

40

The Other Culture*

How can science be presented to non-specialists?

Suppose that 40 per cent of our intelligentsia were accustomed to migrate to China at the age of 15. They make themselves quite at home there, and learn to speak and write the language fluently. Ten years later they return, and enjoy thereafter a sort of dual nationality, spending part of each year in Britain, as ordinary family men and women, but going back to China for weeks or months at a time to take their part, also, as citizens of that country. How could they best explain to the stay-at-homes what it is like to live in China amongst the Chinese?

One can imagine some of the methods that might be tried, and can easily guess their consequences. One group might give a general survey of Chinese geography; this would prove very indigestible and uninspiring. Another institution might concentrate on the history of China, which would be interesting for those with a taste for history, but otherwise a little dry and academic. A very advanced and sophisticated teacher might attempt to discuss the current social problems of the Chinese people, but his listeners would find it difficult to relate this to the simple facts of Chinese society. The attempt to distil a few great thoughts and significant concepts out of Chinese philosophy would fail for lack of understanding of Chinese language; a practical course of how to make a Chinese junk would prove amusing, but not very relevant to the main purpose.

In the end, they would probably discover that the only useful thing to do is to invite their friends and relations to visit China themselves for a short time. They might suggest, for example, a short guided tour of one of the great cities, to be delighted and overwhelmed with the splendour of the ancient monuments and the delicacy of the Chinese cuisine. Others might prefer to spend a few weeks living in some quiet village, sharing the daily life of the peasants, helping perhaps to harvest the rice or to

* Review of *Science for Non-Scientists* by J. S. R. Goodlad (Oxford University Press, 1973) published in *Nature* 242, 579 (1973).

build a new farmhouse. The language barrier would still be an obstacle but with skilled interpreters some modest dialogue might be achieved with the inhabitants; at least the tourists might come to understand that these creatures were human beings and not goblins or mechanical robots, and that they lived in a real country where the grass is green and the sky is blue.

To make the most of such a programme, it would need to be integrated into the education system at all stages. In the primary school, for example, pictures of Chinese landscapes, little games to dramatize the Chinese way of life, and a few elementary lessons on Chinese ideograms would make children sympathetic to China and its people. Throughout secondary school and higher education, this process should continue, so that every educated citizen would be familiar with Chinese ways of thought and action. Unfortunately, there are serious obstacles to this desirable reform. The very sharp segregation of the Sinologues preparatory to their migration, a peculiar feature of the British educational system, is often anticipated by several years, and those who do not emigrate learn practically nothing about China beyond this stage, and regard it as more fabulous than ancient Cathay.

Indeed, many people argue that the division should be much less definite, that those who go should continue to spend part of each of these formative years in their native land and that the non-Sinologues should be made to visit China regularly in school parties or even as individual travellers. All these questions are now widely discussed, but nothing is done; schools and universities blame each other for impeding reform, but no common plan of action has been devised.

At the university level, also, schemes for bridging the gap between the two cultures are being tried out here and there, but without great success. Part of the difficulty is that China itself is a very large country: the Cantonese knows little of Peking, and the Tibetan plateau is very distant from the delta of the Yellow River. Provincial loyalties run high and the honest simple-minded wanderer must find his way through many artificial frontiers. The best results have been obtained with a programme of guided visits to selected areas, but a whole extra year is regarded as too costly by the powers-that-pay. Perhaps insufficient emphasis is laid on the exploration of certain common frontier areas: for, if we study *geography*, we may observe that China begins, in truth, at Calais, and in the eyes of *psychology* all civilizations are one

All these points are made thoughtfully, accurately and with apt citation of many sources, in this excellent book. It is not very long or laborious, nor does it pretend to great subtlety or depth, but it analyzes very shrewdly and positively one of the most serious and difficult problems of our present-day culture.

41

The Examination Society*

The real question is: what are examinations for?

Examinations! Ugh! A dismal thought, especially when the days are long and the sun shines. Pity our golden youth, imprisoned in bleak and graceless halls, scribbling reams of nonsense for dear life, tugging desperately at the curtains that baffle memory and logical thought.

A tear is seldom shed for the examiner. Is he not a mystical monster, descendant of the Sphinx or of Rumpelstiltskin, prepared to rend apart any foolish human who fails to guess his riddles? Who would recognize kind old Professor Bumble, doddery, absent-minded, generous to a fault, under that lion's skin? How *could* he have set such nasty questions; why didn't he *tell* us we ought to swot up Keats, when he had been going on about Shelley all the year?

Poor old Bumble doesn't like examinations either. There are the scripts – a great pile of square packages dumped on his desk by the secretary of the Faculty. Now he must mark them, when he would much rather be composing a learned article, attending a conference by the Adriatic, or dealing death to the greenfly on his roses. 'Let me see; what were the questions now?' – for it is many months since he concocted them, on a busy day in January. Marking is deadly, repetitious labour, requiring concentration and an even temper. 'Fool!', 'Idiot!', 'Nincompoop!', the mind screams silently as the eyes skim over the pages of scrawl – but 'alpha minus', '68 per cent', or 'excellent answer' the red pencil scribbles as the next script is snatched from the pile.

By the time we have finished marking and must attend the examiner's meeting, where the Class List is to be drawn up, we are all hawks. We must firmly defend the standards. We know a First-Class man when we see one – and we don't see one very often. We are appalled at the fantastic errors of the candidates, and sternly vow that such stupidity shall no

* BBC Radio broadcast, 6 July, 1968.

longer be allowed to flourish in the world. 'This year's class is pretty dim', we say, and determine to make it pay for it laziness and folly.

Mercifully, a committee of hawks is as gentle as a dove in action. The icy ferocity of our colleagues germinates the seeds of humanity in our heart. We discover mitigating circumstances: 'He ought to be allowed something for being captain of Chess'; 'One shouldn't really take much notice of the viva, should one – I expect she was just nervous'; 'His tutor speaks quite well of him – perhaps he was a bit off form'; 'Yes, of course, he's not done a stroke of work this year, but we never actually fail anyone at this stage, do we?'

The discussion sways back and forth over the 'borderlines' – those indefinable Tropics that we must somehow draw between the various classes. Tutors' reports are again consulted, and predict with uncanny accuracy that 'He should get a First or at least an Upper Second'. So we 'give the benefit of the doubt' to marginal candidates, and allow the dividing line to fall in the nearest 'natural break' in the distribution of marks; which is just where everybody knew it must go in the first place. Over tea we congratulate ourselves on all the hard work we have done, and on the difficulty of the decisions. Having remarked that it seemed to have been rather a good 'year' after all, we hurriedly sign the official lists, and disperse to more joyful occupations.

Now I don't mean to say that university examinations are not handled honestly and skilfully. We dons are professionals at this game and we do our duty with quite as much care and conscience as any barrister pleading a case, or specialist diagnosing a disease. Our students cover an astonishing range of intellectual power, and of attention to their studies. We are familiar with the inevitable contradiction between the precise standards of scholarship and the confusing haste of a written examination, and we know how to be both hawks and doves as the occasion demands. We do not claim absolute precision in the marks and classes we award, but within reasonable limits the results are reliable and just.

Yet there are moments, over lunch in the Senior Common Room, when another voice may sometimes be heard – that of the cuckoo. 'Down with examinations!' it calls, as if echoing the seasonal protests of disaffected students and their journalistic confederates.

There is more of this than mere anarchy and permissiveness. Exams are not only unpleasant and difficult to pass; they have a very perverse effect upon education. If we take the line of least resistance we fail to teach our students to think for themselves, but just cram them with received opinions, which they regurgitate before us at the end of the year. Despite our best endeavours to convey the deeper truths of our subject, the boys and girls are so fearful of failing that they spend their time learning, parrot-fashion, stereotyped arguments and useless facts. Only by abolishing such artificial incentives can we teach and learn wisely.

And a good examination result is a very crude index of intellectual distinction. True enough, Newton won mathematical scholarships; but Einstein was somewhere amongst the Lower Seconds, and Darwin didn't go for an Honours degree. Even in our own day, when the ladder of academic careerism is so carefully graded, a good proportion of the Fellows of the Royal Society failed to get Firsts as undergraduates. The abilities that are tested in examinations are not trivial; but in later life we must count upon more significant qualities – imagination, perseverance, curiosity, or dexterity. Every teacher is familiar with the student who is lively, thoughtful and original, but who never, as they say, 'does justice' to himself in written papers; so that we are forced to launch him into the world with a miserable Third, as if he were only as dull, stupid or lazy as the worst in his class. When such injustices occur, our hearts echo that cry of the cuckoo, 'Down with examinations!'

Alas, such mutinous thoughts are quickly suppressed by the hooting of the owls that flock in Senior Common Rooms. The owls know very well that there is no substitute for examinations. They will tell you that students are poorly motivated, and cannot be trusted to work steadily for years on end without spur or goad. They recall the fluent, plausible chap, who turned out to have no real depth to his mind, or the quiet one whose intellect shines in the impersonality of a written paper. They can demonstrate the importance and accuracy of relatively objective tests, performed under controlled conditions, freed from the disastrous unreliability and complexity of the face-to-face personal situation. 'Tell me,' they ask, 'have you ever looked at headmasters' reports? Do you really think we could rely upon that sort of information alone, in deciding on admissions? As for interviews, they are completely unreliable; there have even been *experiments* to prove it.' If they have any historical sense, they will go on to explain how corruption and nepotism in the Civil Service and the professions could only have been driven out by a system of competitive examinations. They will point with pride to the multitude of lads from poor homes, with only their intelligence and industry to favour them, who have fought their way up the educational ladder by this means, routing the sons of the rich and influential, and rising to be Colonial Governors, Bishops, Judges, yea even Professors.

The owls will always concede that the *technique* of examinations could be vastly improved. Indeed, they spend an appreciable proportion of their time discussing 'the reform of the Tripos'. They are scornful, for example, of questions that test only a memory for vast quantities of bookwork, so they try to devise more searching problems to discover whether the candidate can 'think for himself'. These problems always turn out far too difficult; in the end the marks have to be scaled up so much as to be quite meaningless. At the moment I detect a ground swell in favour of the multiple-choice, one-word-answer type of question, where the candidate must show his genuine and precise grasp of the subject by picking the correct answer out of a list of plausible-looking alternatives. As an accredited radical in University affairs, of course I approve this change; it is only a little disconcerting that our American colleagues are now moving in the opposite direction, and are extolling the merits of the traditional, British, open-ended, essay-type question, which is so much better at detecting imagination and clarity of expression. Again, it has come to be recognized that such skills as are acquired, for example, in experimental laboratory classes are not adequately assessed by set-piece 'practical' exams, so now the marks are allotted for 'projects' and laboratory notebooks written up during the year; fortunately, these marks are given so little weight in the Final totals that little harm is done one way or the other. 'Continuous assessment', also, has its pitfalls and can become as oppressive as any annual examination festival.

The fact is that the owls are quite right. A conventional written examination, to be completed incommunicado within a limited time, is still about the most reliable test of the sorts of abilities that examinations *can* test at all. If that is what you are aiming at, then this is perhaps the best way to do it.

The real question is: what are examinations *for*? Doves or hawks, owls or cuckoos, we dons seldom try to solve that little riddle.

Nobody in his right mind denies the need for a Driving Test. Nobody nowadays would allow any untrained Tom, Dick or Harry to climb into a motor car and set off alone, at grave risk to himself and others. We insist upon a minimum standard of instruction and performance, which is carefully tested before the issue of a licence. In a society of highly differentiated specialisms, we must also ensure some modest degree of professional competence in the doctors, lawyers, quantity surveyors, architects, engineers, school masters, plumbers – even taxi drivers – to whom we entrust our lives and fortunes. Whether these skills are acquired by formal instruction or by apprenticeship, they must be tested by experts, to protect society from utter incompetence or deliberate fraud. Examinations for this purpose seem to be as essential to our present style of civilization as paved roads and postage stamps.

But notice that the Driving Test, and all its analogues, demands only a minimum level of achievement. It is not an adequate qualification for motor racing, but merely the lowest standard compatible with reasonable safety on the road. We know very well that driving skill improves thereafter for many years, just as we know that a doctor of 35 or 40 is probably much more expert than he was as a raw graduate. With rare exceptions – airline pilots, for example – we do not make our professional licensees pass through a succession of examination hoops, to be sure that they are keeping up to scratch. The normal competition of a career concentrates the mind of most adults quite as well as a periodic ritual hanging; and, as a distinguished French scientist is reputed to have said, when told that he had to submit a thesis for a doctorate before they could make him a full professor: 'Certainly; but who will be my examiner?'

But one can scarcely claim of most university degree examinations that they are necessary to license significant professional skills. A knowledge of Beowulf, or the ability to 'write a washing bill in Babylonic cuneiform' is not particularly relevant in a TV producer, and quantum mechanics is seldom used by school masters, even in the sixth form. No doubt an elementary grasp of history and economics ought to be required of aspiring Members of Parliament; but we have somehow let slip the opportunity to insist upon *that*. Since our system of higher education is in large proportion non-vocational, all these examinations must have some other purpose.

Well, of course, we need them to measure the progress of our education. The pupils must show the teacher explicitly that they have indeed learnt adequately what they are supposed to learn. And it would never do to allow a student to proceed to next year's course unless he could demonstrate adequate understanding of what he had just been subjected to. Why, even the student himself will surely want to be tested just to find out how he is getting on.

This is legitimate – but I don't see that it requires all the rigmarole of five written papers, printed by the University Press and later published at the price of 3/6 per set, to discover that sort of thing. In Britain, anyway, once we have admitted students to the University, we seldom allow them actually to *fail* a course of study – and then we inquire into the circumstances, and switch on all the tutorial care circuits in the hope that the real trouble is an unhappy love affair, or an incipient nervous breakdown, or slowness to adjust to the University atmosphere, rather than primitive laziness or stupidity. With only an average of a dozen or so students per teacher, we ought to be able to watch their progress quite easily, by reading their essays, supervising their practical work, helping them solve problems, even by just talking to them from time to time throughout the session.

Or why not make progress examinations voluntary? Let them be set now and then by the instructor, and let any student who wishes try his hand at them, for his own edification. Let the University offer various courses of study, at various levels of difficulty, to be 'passed' or not, just as one likes. Why should a young man be deemed to have 'wasted his time' if he does not sit the Final Degree examinations and receive the appropriate certificate? The doctors and engineers must, of course, be licensed to practice; but for the would-be management trainee or advertising copywriter, it is his own business to decide whether he should submit his views on *Finnegans Wake* to such unenlightened comment, or interrupt his interesting reading of the latest trends in sociology in order to 'compare and contrast Napoleon with Louis Quatorze'. The whole function of non-vocational education would surely be enhanced by removing these futile constraints.

To this the owl in me has a ready answer; you must justify to the State the money that is being spent on running the Universities and maintaining the students. The progress of every student must be carefully measured and charted, or else public money would be wasted. Even if few are actually 'weeded out' (oh monstrous metaphor!) the annual threat is held to be essential to their moral welfare and intellectual industry.

Very well then, I'll go further. Let us replace student grants by personal loans, on easy terms, and raise fees to cover the whole real cost of higher education. University staff would no longer have to report back on the progress of every pupil. Each student would have his own financial incentives, whether to get on and out into the world or to take the risk of a further year's study, with its likely spiritual and material benefits. By taking upon himself the responsibility for such decisions, he would free himself from the terrible burden of compulsory examinations as we know them.

The system I am sketching is not wildly impracticable; but it also has its disadvantages, as anyone will know who has experienced something like it on the Continent or in the United States. And I had better say no more on such a vexed subject, for I might find a picket of the Student Union occupying my study one fine morning! But if we are to think about examinations at all, outside of the conventional wisdom, we must recognize their connexion with the other customs of our society; and perhaps even dare to imagine that these too might be altered.

The heart of the matter is that we have got into the habit, during the past century, of using examination results as the arbiters of personal fate. A progress examination that detects a few members of the class in grave difficulties is a humane instrument; a qualifying test that excludes the grossly unfit from a specialized, responsible office is a necessary evil; but the mechanical use of written examinations to shut off permanently the life aspirations of vast numbers of moderately competent individuals is a monstrosity. The idea that any sort of test, whether at 11-plus, at 18-plus, or at 22, could be an infallible index of inherent merit, or of vocational aptitude, valid thereafter for life, is a gross fallacy.

My colleagues and I are perfectly competent to report that a particular script indicates only an indifferent comprehension of our subject, and may award an appropriate, approximate, conventional mark. We may properly advise the student not to attempt the next stage of study in this subject without further preparation. We may even rule that with only 50 vacant places in the course, those with the highest marks should have preference, and we can insist that the weak candidate should seek to further his education elsewhere. But we are *not* thereafter entitled to refer to him as 'only a Third', as if to damn him for ever. Shades of *Brave New World:* 'I'm glad I'm not a Gamma!'

Examinations are worse than tiresome obstacles along the road – they are positively immoral – if their results are taken seriously. They stand for an achievement here and now, not for eternity.

I know that the academic professions are the worst offenders. Consider, for example, that rule by which a graduate with Second-Class Honours gets an extra £120 a year, *for life,* when he becomes a teacher. Or a recent occasion when we were discussing a certain lecturer in a College of Education. 'Yes, yes, I know he's an excellent teacher; but look at his degree results; he only got a Third!' Good grief, that degree had been taken *30 years before.* We dons need to be reminded of that humbling truth: 'There are no failed students, only failed teachers!'

But do not blame us utterly, for we are only the agents of a general social force. The Civil Service may begin to reform itself, but the same fallacious rigidity of mind is to be found throughout the bureaucratic world. The situation is just as bad elsewhere; at Harvard they try pep pills to improve their grade point average; in Delhi, the students just go on strike. Competitive society will seize on any instrument of differentiation, however bizarre, for its irresistible purposes.

Alas, exams, like death and taxes, are with us for a long time yet; but I have a simple proposal for a saner world. Let it be enacted that all examination results shall henceforth be recorded only in fading ink, on crumbling paper, in damp, locked, filing cabinets; and let it be an offence tantamount to libel to mention any such result that is more than five years old.

Well, anyway, with the long vacation before us, let's try to forget the whole sorry business for another year.

42

Some Manifestations of Scientism*

Teaching about science should lay particular stress on the limitations of the scientific attitude.

Science is not easily incorporated into human affairs. It is so dynamic, it transforms our material culture so dramatically, that we have not yet found for it a stable role in our social system. Science has not got an established place like the Law or the Church; the general public and their representatives do not know how to treat this forceful, ever-growing intruder on the political scene. One of the tendencies of our times is to exaggerate this role and to treat science as the dominant actor in social affairs. This tendency is called *scientism*. The essence of scientism is to regard science, not merely as a very active force, but also as somewhat more efficacious than it really is. This could be as grave a folly as putting one's head into the sand and wishing science would go away, with all its works.

Scientism has a long history. In the period of the 'Enlightenment' of the eighteenth century, many people came to believe in the beneficial power of a rational approach to human problems, which they thought might be solved with the same sweeping success as Newton had solved the problems of astronomy. The Industrial Revolution gave rise to confidence in the material progress being achieved by the application of science and scientific technology. This was not merely the optimism and complacency of a comfortable middle class, being rapidly enriched by manufacturing industry; even revolutionary socialists such as Marx and Lenin took for granted that scientific progress was capable of making the world better in every way, and that knowledge only had to be applied properly to benefit mankind.

Of course, there were always those who reacted strongly against this point of view. The Romantic movement of the early nineteenth century established a tradition of cultural 'anti-scientism' represented by writers

* Background paper written in 1978 for the Science in Society Project of the Association for Science Education.

such as Matthew Arnold in the late nineteenth century, C. S. Lewis in the 1930s and 1940s, and Theodore Roszak in recent years. These works emphasize aesthetic and spiritual values which lie outside the ken of scientific knowledge, but often fall into the scientistic error of making science seem more powerful, for inhumanity, ugliness and evil, than it really is. This extreme opposition to science was, until recently, confined to quite narrow, highly intellectual circles and had very little influence on general popular culture, where science came to be regarded as a sophisticated variety of magic from which great benefits could be expected to flow.

This naïve form of scientism is still predominant in most people's minds. During the last decade, scientists have indulged in an orgy of self-criticism and confessions of 'having known sin'; yet opinion polls in all advanced industrial countries show that there is still enormous confidence in the capacities of scientists and in the benevolence of their work. But thoughtful people no longer assume that the mere accumulation of scientific knowledge and of technological novelty will automatically put the world to rights. There is grave suspicion of material progress as a path to happiness and salvation.

There are doubts, also, about the wisdom of trying to cure our economic and social ills by a deliberate 'technological fix'. Twenty years ago, for example, there was great enthusiasm for enormous nuclear power projects that would provide electric power, desalinated sea-water for irrigation, and nitrogenous fertilizers, to make the deserts bloom. Traditional poverty and social misery were thus to be eliminated in one bold plan. Nowadays, that sort of scheme is regarded as impossibly grandiose and almost certainly pernicious in its consequences. A modest degree of technological optimism is fully justified by our experience and understanding of the capabilities of modern engineering, agriculture and medicine; but there are few who believe that the way to a better life for all mankind is just to let rip with more and more research and more and more sophisticated technical tricks.

Naïve scientism is in decline; but undue deference to science is still the predominant theme of our culture. Some of the manifestations of this theme are positive, in that science is regarded as highly efficacious for good; others are negative, asserting that science is inevitably tainted with inhumanity and evil. The opposing parties are unwittingly in agreement on the fundamental scientistic doctrine that science is a coherent entity which must either be harnessed for good ends or else put in chains to prevent the ruin of the world.

Academic scientism pervades the intellectual world. Every academic discipline is squeezed into a 'scientific' mould. Not only do we have 'Materials Science' and 'Medical Science' but also 'the Social Sciences' and even 'Political Science'. In some languages the word that we translate in English as 'science' has a broader meaning. The German word, *Wissenschaft,* for example, can be applied to any well-organized body of knowledge such as History or Literary Criticism. But those who give the English name 'Political Science' to the study of political institutions and of political theory are quite deliberately asserting that their scholarly activities are closely akin to the styles of teaching and research prevalent in a department of biochemistry, or geology. The implication is that by the use of quantitative methods, theoretical models, observational data, experimental techniques, etc., they are well on the way to achieving the same powers of prediction and technological application, in the sphere of politics, as physics, say, has achieved in the sphere of mechanical engineering.

In its simplest form, this aspect of scientism derives from the philosophy of *reductionism*. Since societies are made up of people, people are biological organisms, biological organisms are made up of cells, cells are complex chemical assemblies, a chemical molecule is made up of atoms, atoms consist of nucleons and electrons (etc., etc.), then if we really understand all about electrons and nucleons by doing physics, we shall, eventually, be in a position to understand the behaviour of human society. This doctrine is of limited use in practice, because it is very difficult to demonstrate the logical necessity of hypothetical connexions between scientific knowledge at one level, for example, molecular biology, and a property of a more complex system, such as a cell or an organism. But reductionism is still very much alive as a basic metaphysical principle, within science itself and amongst many of those who admire its achievement.

Philosophical scientism is the point of view that only 'scientific' statements should be regarded as really truly true. Thus, for example, a study of political phenomena should confine itself to the sorts of statements that are characteristic of scientific facts and theories, and these statements should be analysed according to the methods that have proved to be successful in science. This is the view technically known as *positivism*. It is not new in Western philosophy; it probably reached its peak in the Logical Positivism of the 1930s and is now on the wane. But it is still widely held in the academic world as an unspoken principle in the formulation or criticism of research programmes, even in fields outside the traditional natural sciences. Thoughtful scholars are well able to see the dangers and deficiencies of such a narrow point of view within their own particular fields. Positivism is not only discredited philosophically; in every discipline there is great debate about the appropriateness and validity of quasi-scientific methodologies and theories. But the effort to model every form of organized knowledge on the natural sciences has certainly not been abandoned, and still remains an unconscious motive throughout the academic world.

Practical scientism manifests itself in a belief in the superiority of what is called the 'scientific attitude' in active life. This phrase was very prevalent before the Second World War – for example, in the popular writings of Bertrand Russell. This attitude combined ethical and rational virtues supposedly derived from the experience of scientific research: thus, for example, because scientific knowledge is universal and international, the scientific attitude must be for peace and against war; because science is sceptical and analytical, most of the problems of the world could be solved by the application of sceptical, analytical thought. In its most extreme form, this idea is exemplified in Fred Hoyle's science fiction romance *The Black Cloud,* where the world is saved by a Professor of Astronomy who unites in his person all those ideal qualities of character and intellect that are supposed to be typical of the research scientist.

The all-round virtue of the scientific attitude has lost some of its credibility in recent years. There is growing disenchantment with the ethical aspects of the practice of research. Books describing the realities of scientific life, such as *The Double Helix,* have dispelled illusions about the high moral stature of scientists, even within their own laboratories. Nobody now supposes that a lifelong pursuit of theoretical physics or neural physiology automatically gives one a special insight into what it is good or bad to do in practical life.

But belief in the efficacy of the 'scientific method' for solving practical problems dies hard. For example, the enthusiasm for Systems Analysis as a decision-making tool in fields as diverse as economics, ecology, production engineering, vote-getting, energy policy, etc., etc., is a typical manifestation of scientism. The electronic computer, the epitome of the logical, rational, quantitative, scientific instrument, is applied to every rationalizable aspect of life, regardless of whether any such application is justified. Pessimistic futurologists try to take more account of the diversity of material and human factors in the real world than the optimistic technological fixers of a previous generation; but their claims to be able to predict the future in any detail must be treated very sceptically. There is no evidence from actual experience that the behaviour of a complex biological organism or society of organisms can be represented by a mathematical model, and reliably predicted over any significant period of time.

But even when these balloons have been pricked, scientism remains cheerfully alive. Science is evidently a tremendous influence on society; it follows that scientists are tremendously influential people and must exert this influence for the best. Yes, we realise that the practice of science does not make people wise or good in practical affairs; but, surely, these deficiencies could be remedied by suitable education.

The movement for 'social responsibility in science' is a proper and long-overdue reaction to a tradition of complete professional specialization in research, without regard to its political, social or economic context. But the phrase carries an implication that the scientist has a peculiar right or duty to act within the social sphere whenever science is being applied. There is an unvoiced assumption that scientists ought to be ready to take on a degree of social power that goes beyond their ordinary rights as citizens, or their duties, as experts, to say what they know. An explicit linkage is made between training in the contents of science, or the skills of scientific research, and an understanding of all the practical, political, economic, cultural and social problems of the world – war, overpopulation, racialism, poverty, pollution, 'quality of life', etc. A properly educated scientist is supposed not only to understand what science is and how it works; he or she is apparently being prepared for a much grander social role, in which it will be necessary to grasp the basic elements of economic theory, biomedical practice, agriculture, industrial management, etc., and by special training in the arts of 'decision making', to know how to apply this knowledge in the corridors of power. In other words, scientism again becomes manifest as belief in the benefits of *technocracy* – government or management by technical experts.

It is a very good thing that people should be educated formally or informally towards some grasp of the great problems of our time. Political, economical, social and cultural issues should not only be commented on in the newspapers or made terrifyingly vivid on TV; they need to be learnt about and discussed in the classroom, around the family dinner table and in the pub. But it is not only the specialist or would-be specialist in the natural sciences who needs to be made aware of these issues. In a pluralistic, democratic society it is just as important that the expert in the Fine Arts, Literature, Religion, Law, Social Welfare, Industrial Management, Political Administration, etc., should have a broad education and some confidence that he too would have some contribution to make to the solution of the problems of the world or of the nation. By over-emphasizing the 'relevance' of the natural sciences and technology, ignoring alternative linkages, such as 'the arts and society', 'the humanities and society', 'history and society', 'psychopathology and society', 'literature and society', 'religion and society', 'philosophy and society' or 'language and society', we are liable to fall once more into the scientistic trap from which we thought we were trying to escape.

The real danger of scientism is that it fails to recognize the relatively specialized role of science, and of scientists, in society. Without minimizing the importance and power of science in its own spheres, we need to be much more realistic about the limitations of those powers. These limitations are evident in many different dimensions:

(*a*) Scientific knowledge is reliable only over certain aspects of the natural world – principally, those aspects that are studied in the physical and biological sciences. And even in these traditional disciplines, many valid questions remain unanswered, or are tacitly ignored, because they do not seem to come within the grasp of 'the art of the soluble'.

(*b*) Even within the traditional natural sciences, it is far more difficult to arrive at a reliable scientific answer than is imagined by those who know only about the great triumphs of research and discovery. For every successful investigation there are dozens that fail to reach a convincing conclusion. The main principles of physics, chemistry and biology may well be satisfactorily understood; but the work of elucidating the details of all the phenomena that are supposed to be governed by these principles is lengthy, laborious and grotesquely incomplete. Much of our science-based technology works well enough, not because it was designed 'scientifically' from first principles with a real understanding of what was going on, but simply because it has been tested in practice by old-fashioned methods of trial and error.

(*c*) In many spheres of rational knowledge, dealing with many observable aspects of the natural world – especially the individual and social behaviour of biological organisms – there is an almost complete lack of reliable, fundamental knowledge. Many particular facts are known, but there is inadequate evidence to support and make fully convincing a general set of rigorous principles analogous to the 'laws' of the physical sciences. Every claim to have discovered such overriding principles (from which strictly verifiable predictions would follow) must be regarded with complete scepticism.

(*d*) Success in the technological applications of the physical sciences and applied mathematics does not justify a belief that the behaviour of any complex, strongly interacting system can be accurately predicted over a long period. Quantitative data processing, an understanding of underlying processes, and the capacity for strict logical deduction, are all helpful in the control of any complex system, such as a spacecraft, a chemical plant or a national budget. But such control cannot be maintained without continual monitoring, and unforeseen crises can only be handled by the light of experience and good sense.

(*e*) Political decision-making, and every other form of social action, is constrained by imperfect knowledge of the situation, the short time available for cogitation and the multitudinous possibilities of wickedness and folly. Success depends on a diversity of skills and insight such as practical experience, moral rectitude, empathy, honesty, patience, idealism, cunning, charismatic authority, etc. These talents usually prove far more effective (for good as well as ill!) than the best knowledge available to science.

(*f*) The career of the research scientist seldom includes situations where rapid decisions must be taken under conditions of uncertainty, moral ambivalence, or the conflict of irreconcilable interests. For this reason scientists are amongst those persons in society whose experience ill prepares them for the most demanding responsibilities of politics, business or war. Although some scientists do, in fact, show these talents in practice, there is no case for forcing this social role onto all scientists simply because they are engaged in producing socially powerful knowledge.

The social function of science is more diffuse, and more complicated, than is allowed for in scientistic ideologies. To a first approximation, scientific knowledge is no more than a public archive which may be consulted for useful or harmful ends, as people decide. The scientist is no more than a professional specialist in the generation of new contributions to this archive, or in the synthesis of what is already known. His proper role is that of any technical expert, whose opinions will be sought, when necessary, by the powers that be. As one politician put it: 'Scientists should be on tap, not on top.'

But this traditional point of view ignores the extent to which science, in a variety of forms, permeates our culture. Scientific discoveries, concepts and methods profoundly alter our image of the world, and deeply influence all other branches of scholarship and learning. Through industrial, medical, agricultural and other technologies, science is transforming the manner of our lives. There is an immense demand for reliable scientific knowledge to be applied for private profit or public benefit, and enormous sums are spent on research projects with very practical ends.

The scientists themselves continually clamour for expensive research facilities to study fundamental problems without any specific application.

Science can no longer be regarded as a realm apart from the everyday world, a distinct sub-culture sufficient unto itself. And the scientist who regards himself as set apart, and entirely committed to research in a very narrow field – as, so to speak, a mere embodiment of a specialized subdivision of the scientific archive – can no longer make a satisfactory contribution to the world by his work. By nature, or by nurture, he may have immensely valuable capabilities, such as insatiable curiosity, obsessive concentration, probing logic, unwearied patience, or unfettered imagination. But for the most profound contributions to knowledge, he needs to see his own subject in a broader intellectual setting, in the context of other academic disciplines. To make credible claims for funds and facilities for research, he needs to be fully aware of its potentialities for application. To preserve his moral self-esteem, he needs to be alert to the motives of those who would seek to exploit his skills and knowledge. To give of his best as an expert adviser, he needs some grasp of the political, social and economic realities within which that advice is to be acted upon. And if he is to be effective in a political or administrative role, he needs a well-founded respect for the skills of other experts whose life experience has been very different from his own.

At the same time, the non-scientist needs a realistic understanding of the capabilities of science. He needs a feel for the validity of its mysterious concepts and theories. He needs to know what technological applications are being contemplated, and the benefits or harm they might create. He needs a little experience of the power of scientific methods of problem-solving, and adequate appreciation of their limitations. Above all, he needs sympathy with research as a creative and beneficial vocation – as something more, to those involved in it, than a professional craft.

The social role of science is not yet fixed; to a large extent, it depends on what use we make of it. But that, in turn, depends on a just appreciation of what science is, and what scientists can do. The folly of scientism is to lump together and exaggerate a diversity of tendencies, forces and motives. Science is *not* an irresistible force, sweeping like a tide over human affairs. It is *not* necessary for us all to acquiesce to the introduction of any technological innovation that any rascal can sell to any fool. It is *not* inevitable that science will fall passively into the hands of an all-powerful state. All knowledge is *not* scientific. All human problems are *not* soluble by the application of scientific method. Scientists are *not* all villainous or all divine. The pursuit of knowledge is *not* an uncovenanted good. And so on, and so on. Acceptance of any one of these extreme attitudes makes people blind to the realities of a changing world, so that they miss many opportunities for positive action for good.

But the most damaging consequence of scientism is that it creates barriers of pride, envy, arrogance and prejudice between the most dedicated and competent members of society. Superficial contrasts and antagonism between 'the sciences' and 'the humanities' disrupt our culture into disconnected sub-cultures, served by mutually suspicious, mutually uncomprehending technical experts. The issues on which we seek enlightenment cannot be seen clearly except by the harmonious combination of many different points of view. The problems of our times are quite difficult enough; they will certainly never be solved unless there is respect and open communication between scientists and non-scientists as they work together in practical affairs.

INDEX